AF598100

# CARBOXYLIC ACID - KEY ROLE IN LIFE SCIENCES

Edited by **Georgiana Ileana Badea**
and **Gabriel Lucian Radu**

**Carboxylic Acid - Key Role in Life Sciences**
http://dx.doi.org/10.5772/intechopen.68350
Edited by Georgiana Ileana Badea and Gabriel Lucian Radu

**Contributors**

Saša Đurović, Saša Šorgić, Saša Popov, Zoran Zeković, Marija Radojković, Jörg Gerke, Aide Saenz, Lluvia I López-López, Fabiola N. De La Cruz, Adali O. Castañeda, Leticia A. Ramirez, Karla C. Cordova, Denisse De Loera, Kouichi Matsumoto, Yohei Oe, Rina Yanagi, Georgiana-Ileana Badea



First published in London, United Kingdom, 2018 by IntechOpen
IntechOpen is the global imprint of INTECHOPEN LIMITED, registered in England and Wales, registration number: 11086078, The Shard, 25th floor, 32 London Bridge Street
London, SE19SG – United Kingdom
Printed in Croatia

British Library Cataloguing-in-Publication Data
A catalogue record for this book is available from the British Library

Additional hard copies can be obtained from orders@intechopen.com

Carboxylic Acid - Key Role in Life Sciences, Edited by Georgiana Ileana Badea and Gabriel Lucian Radu
p. cm.
Print ISBN 978-1-78923-278-3
Online ISBN 978-1-78923-279-0

For EU product safety concerns:
IN TECH d.o.o., Prolaz Marije Krucifikse Kozulić 3, 51000 Rijeka, Croatia,
info@intechopen.com or visit our website at intechopen.com.

# Meet the editors

Dr. Georgiana Ileana Badea earned her BS degree (2007) in Chemistry and MSc degree in Optimisation of Chemical-Analytical Control and Quality Assurance (2009) from the University of Bucharest, Faculty of Chemistry, Bucharest (Romania). She received her PhD degree (2012) in Chemistry from the Polytechnic University of Bucharest, Faculty of Applied Chemistry and Material Sciences. Since 2007, she has been employed as a scientific researcher at Center of Bioanalysis, NIRDBS, Bucharest, Romania. Her research work involves the development of analytical methods for identification and quantification of biological active compounds, methods for authentication of archaeological findings, physico-chemical characterization of biomaterials (nanoparticles), release of targeted compound studies, etc. She has written over 20 scientific publications and 3 book chapters in the aforementioned areas.

Prof. Dr. Radu Gabriel-Lucian is a senior researcher at the Center of Bioanalysis, NIRDBS, Bucharest, and Professor at the Department of Analytical Chemistry and Environmental Engineering at the Polytechnic University of Bucharest. His research/teaching activity focus has been in the areas of life sciences, in particular on the development of new analytical methods for characterizing bio-compounds and bio-processes. He co-ordinated more than 55 national and international projects. He is a PhD research supervisor in the Chemistry field since 1998 with 24 doctoral theses so far being supported under his scientific leadership. He has published over 177 peer-reviewed scientific publications in international journals and served on the board of many Scientific Associations and Committees.

# Contents

# Preface

This book is an attempt to bring together current knowledge on the role and importance of organic acids in life processes. No overall consensus is sought in this book, and the following chapters are authored by dedicated researchers presenting a diversity of applications and hypotheses concerning organic acids.

The five chapters in this book include general information regarding carboxylic acid properties, the importance and their applications in life sciences (use in organic synthesis, nanotechnology, fields of plant physiology, plant nutrition and soil chemistry) and one case study regarding isolation and analytical quantification of a particular class of organic acids (fatty acids) from stinging nettle leaves.

The introductory chapter provides general information regarding this class of compounds starting with physico-chemical description and closing with the significant domains where the presence and the actions of these carboxylic acids have a great impact.

The second chapter addresses the recent advances in the synthesis of carboxylic acid esters based on utilization of carboxylic acids as electrophiles or nucleophiles in reactions. In this chapter, several reactions classified based on the use of carboxylic acids as (1) electrophiles or (2) nucleophiles are described applying various chemical reagents and catalysts.

The third chapter provides a comprehensive overview on the use of carboxylic acids in different areas such as organic synthesis, nanotechnology and polymers. Carboxylic acids are used as surface modifiers in carbon-based nanostructures: carbon nanotubes, nanofibers with the purpose of improving the solubility in polar solvents. They also have significant applications in the polymer field having different functions such as monomers and additives or used as initiators, catalysts or dopants.

The fourth chapter offers detailed discussion of mechanisms of organic acid on the acquisition of soil phosphate in the fields of plant physiology, plant nutrition and soil chemistry. Some plant species strongly mobilize soil phosphate by carboxylates improving this macronutrient acquisition.

In fifth chapter, a research study whose main goal was to isolate and establish the fatty acid profile in the leaves of wild growing plants from Urticaceae botanical family (*Urtica dioica L.*, stinging nettle) is presented.

By presenting a few aspects regarding the role and importance of this class of compounds in key life cycles, it is expected to stimulate the interest of more scientists to dedicate their researches in highlighting and/or improving the mechanism/strategies of action and analyses of carboxylic acids and their derivatives.

**Georgiana Ileana Badea**
National Institute of Research and Development for Biological Sciences, Romania

**Gabriel Lucian Radu**
Polytechnic University of Bucharest, Romania

Chapter 1

# Introductory Chapter: Carboxylic Acids - Key Role in Life Sciences

Georgiana Ileana Badea and Gabriel Lucian Radu

Additional information is available at the end of the chapter

http://dx.doi.org/10.5772/intechopen.77021

## 1. Introduction

Carboxylic acids or organic acids are the compounds containing in the molecule the carboxyl functional group attached to the hydrocarbon radical. They are largely distributed in nature and are intermediates in the degradation pathways of amino acids, fats, and carbohydrates.

The carboxyl group consisting of a carbonyl (C=O) with a hydroxyl group (O–H) attached to the same carbon atom and is usually written as –COOH or CO2H. The compounds presenting two or more carboxylic groups are called dicarboxylic, tricarboxylic acids, while their salts and esters are called carboxylates. By the nature of the radical, they can be classified into saturated, unsaturated, or aromatic acids. In the International Union of Pure and Applied Chemistry (IUPAC) nomenclature, carboxylic acids have an "-oic acid" suffix added to hydrocarbons having the same number of carbon atoms. Still, some organic acids are called by their common name, for example, formic acid and acetic acid.

The molecular weight of organic acids varies widely from relatively small compounds such as formic and acetic acids too much larger compounds (fatty acids) with higher numbers of carboxylic and phenolic functional groups. Monocarboxylic acids with 5–10 carbon atoms in the chain are colorless liquids with unpleasant smells. As the carbon chain length increases (>10 carbon atoms) the acids are waxlike solids, and their smell diminishes with increasing molar mass and decreasing volatility.

Organic acids are weak acids with pKa values ranging from 3 (carboxylic) to 9 (phenolic) meaning that they do not dissociate totally in a neutral aqueous solution to produce H+ cations. The representative low molecular weight organic acids (formic, oxalic, and malic) have a relatively low pKa (<4.0).

Due to the presence of both hydroxyl and the carbonyl groups in the molecule, the carboxylic acids can exhibit hydrogen bonding with themselves leading to increased stabilization of the compounds and show elevated boiling points. They are polar molecules soluble in polar solvents, but as the alkyl chain increases their solubility decreases due to the hydrophobic nature of the carbon chain. In non-polar media, carboxylic acids exist as dimeric pairs due to their capacity to form hydrogen bonds [1].

## 2. Applications in Life Sciences

Carboxylic acids are compounds occurring naturally in different stages of life cycles (living organism-Krebs cycle; fermentation processes, and geological processes) or can be produced in the laboratories or at large scale (synthesis) from oxidation reactions of aldehydes, primary alcohols, and hydrocarbons, oxidative cleavage of olefins, base catalyzed dehydrogenation of alcohols or through the hydrolysis of nitriles, esters, or amides. The organic acids play significant and varied roles in our contemporary society as evidenced by multiple applications in the field of medicine, agriculture, pharmaceuticals, food, and other industries.

Carboxylic acids and their derivatives are used in the production of polymers, biopolymers, coatings, adhesives, and pharmaceutical drugs. They also can be used as solvents, food additives, antimicrobials, and flavorings.

Organic acids have important roles in the **food industry,** since they affect the organoleptic properties (e.g. taste, aroma, and color) and the stability of food items. They can be present as natural food components, for example, the acids present in fruits and vegetables (citric acid in citric fruits, malic acid in grapes and apples, oxalic acid salts in parsley, broccoli), or added artificially, as acidulants (citric acid), preservatives (lactic acid), emulsifiers (tartaric acid), antioxidants (ascorbic acid), or flavors (propionic acid) in a wide variety of products for human consumption (foods and beverages). The level and nature of organic acids present in foods and drinks provide relevant information for monitoring the fermentation processes, control the production, storage, and distribution stages or identify possible adulteration actions. Precisely for this purpose, analytical methods need to be continuously developed and applied in order to identify and quantify the amounts of different acids present in food and beverages. The beverage industry (juices and drinks) is one of the most controlled and regulated industries in terms of composition and authenticity of products. Organic acids are well-known as effective preservatives, and their antimicrobial action is due to the ability to change from undissociated to dissociated form, depending on the environmental pH, making them effective antimicrobial agents. An example, some organic salts (calcium and sodium propionate) prevent spoilage by inhibiting the growth of bacteria and fungi and are used as preservatives in dairy and bakery food products. However, there are carboxylic acids that have a beneficial effect on micro-organisms, helping their growth by acting as vitamins for microbial nutrition (e.g. folic acid, nicotinic acid, or p-aminobenzoic acid). Several studies on the inhibitory effect of various organic acids (oxalic, citric, and malic acids) on polyphenol oxidase (PPO), the enzyme responsible for the browning of damaged fruits and vegetables, have been conducted over the years. The successful results of these studies have had a positive economic impact in the food industry, where maintaining the quality and extending the products shelf-life represents a necessity [2].

Carboxylic acids also play significant roles in the **medicinal fields**. Since organic acids are intermediate metabolites of all major groups of organic cellular components, it has been repeatedly proven that their presence in excess in various fluids of the human body is linked to the manifestation of certain diseases. Organic acids are indicators of organic acidurias associated with various inborn errors of protein metabolism. More than 65 disorders well-known these days are due to enzyme deficiency in the amino acids degradation pathways (leucine, isoleucine, valine, homocysteine, tyrosine, methionine, threonine, lysine, and tryptophan) resulting in an increase of the organic acid concentration in circulation or excreted urine [3]. This toxic accumulation of metabolites, which are not present under physiological conditions in the organism, causes an intoxication-like clinical condition. The urinary organic acid pattern yielded from these metabolic abnormalities is essential for diagnosis. For example, the levels of homovanillic acid (HVA) and vanillylmandelic acids (VMAs) in body fluids are used in the diagnosis of neurological diseases and disorders. Studies of the metabolic fingerprints associate the alternative less efficient degradation pathways of fatty acids (leading to increased levels of adipic and suberic acids in urinary excretion) with disorders like autism. The levels of succinic acid in clinical samples indicate the occurrence of a bacterial infection without the possibility of differentiation between aerobic and anaerobic bacteria type. The quantification of organic acids levels in body fluids can provide useful information in critical areas of metabolism: neurotransmitter metabolism, gastrointestinal function, cellular energy and mitochondrial metabolism, and amino acid/organic acid balance with the purpose of an early diagnosis of various diseases.

The **pharmaceutical industry** benefits as well from the presence and the functions of carboxylic acids. Explaining the importance of carboxylic acids and their derivatives in the pharmaceutical industry rely on the chemical nature of the functional group. The most important roles that carboxylic functions play in pharmaceuticals are:

- Solubilizer acting in modulating solubility, lipophilicity, and cell permeation (e.g. antibiotic or antihistaminic drug classes);
- Prodrug and/or bioprecursor acting as compounds not biologically active but converted into active ones in specific conditions (e.g. drugs from antihypertensive, antithrombotic, or antiviral classes);
- Pharmacophore providing specific interactions with an enzyme, triggering, or blocking its biological response (e.g. blood cholesterol-reducing drugs, nonsteroidal anti-inflammatory drugs).

Carboxylic acid-containing drugs play a major role in the medical treatment of pain and diseases [4].

They are also used in a wide variety of applications as ingredients in cosmetics. A class of organic acids with an important contribution in the cosmetic field is the alpha hydroxy acids (AHAs).

Citric, malic, tartaric, and lactic and glycolic acids are part of this category and are extensively used in cosmetics for purposes such unblock/clean pores, improve the skin texture, whitening, anti-wrinkle, or acne treatment. Also, carboxylic acids represented by aldobionic acids (ABAs), retinoic acids, vitamin C, and azelaic acid are most effective in providing antioxidant and anti-aging protection, as well as improving moisture-retention [5, 6]. The carboxylic acid-based esters are the derivatives most well-known for their flavors and fragrances and are widely used in various applications including perfumes, deodorant, and air fresheners.

Fatty acids represent the class of carboxylic acids recognized for its utility in the cosmetic industry since their water-soluble salts (soaps) have been used as cleansers, since antiquity and are the most useful surfactants known.

Although is a controversy issue about the role of organic acids in **geological processes**, a collection of studies and data provide insights and research directions concerning their influence on geological processes. Many reports suggested that organic acids have participated in soil formation, surface weathering, subsurface porosity generation, and ore formation. Also, studies described the significant and varied roles of organic acids played in rhizosphere acidification and mineral weathering. Literature data list the important roles of organic acids in rhizosphere processes such as mobilization of nutrients (Fe, phosphates), protection against aluminum toxicity, increasing the weathering rate of primary minerals, and participating in translocation of Fe and Al. Other classes of carboxylic acids such as phenolic and fatty acids inhibit the normal growth of plants and algae, acting as allelochemicals, resulting in adverse impacts on the physiological and ecological environment (changed microflora) [7].

One chapter of this book offers, the detailed discussion of mechanisms of organic acid on the acquisition of soil phosphate in the fields of plant physiology, plant nutrition, and soil chemistry. Some plant species strongly mobilize soil phosphate by carboxylates improving this macronutrient acquisition.

## 3. Conclusions

The carboxylic acid compounds still may find applications that cannot be fully covered in this chapter. As conclusion, starting from food to medicine, from the human body to earth and environment, the production, destruction, absorption, or release of these compounds show a strong impact on all the processes/reactions that take place.

As a final conclusion, this subject is an endless one and the classes of compounds that contain the carboxyl functional group, along with all their derivatives, are inseparable from everything that life means on this earth.

## Conflict of interest

Replace the entirety of this text with the "conflict of interest" declaration.

## Author details

Georgiana Ileana Badea[1]* and Gabriel Lucian Radu[2]

*Address all correspondence to: truicageorgiana@yahoo.com

1 National Institute of Research and Development for Biological Sciences, Bucharest, Romania

2 Polytechnic University of Bucharest, Bucharest, Romania

## References

[1] Voet D, Voet JG, editors. Biochimie. 2nd ed. Paris: De Boeck Universite; 1998. p. 1361

[2] Lei Zhou, Wei Liu, Zhiqiang Xiong, Liqiang Zou, Jun Chen, Junping Liu, Junzhen Zhong: Different modes of inhibition for organic acids on polyphenoloxidase. Food Chemistry. 2016;**199**:439-446. DOI: http://dx.doi.org/10.1016/j.foodchem.2015.12.034

[3] Chalmers R. Organic Acids in Man. 1st ed. Netherlands: Springer; 1982. DOI: 10.1007/978-94-009-5778-7

[4] Kalgutkar AS, Daniels JS. Carboxylic Acids and their Bioisosteres. In: Smith DA, editor. Metabolism, Pharmacokinetics, and Toxicity of Functional Groups: Impact of the Building Blocks of Medicinal Chemistry on ADMET. London: Royal Society of Chemistry; 2010. pp. 99-167

[5] Yu R, Eugene J. Alpha-hydroxyacids and carboxylic acids. Journal of Cosmetic Dermatology. 2004;**3**(2):76-87. DOI: 10.1111/j.1473-2130.2004.00059.x

[6] Lukic M, Pantelic I, Savic S. An overview of novel surfactants for formulation of cosmetics with certain emphasis on acidic active substances. Tenside Surfactants Detergents. 2016;**53**(1):7-19. DOI: 10.3139/113.110405

[7] Xiao M, Wu F. A review of environmental characteristics and effects of low-molecular-weight organic acids in the surface ecosystem. Journal of Environmental Sciences. 2014;**26**:935-954. DOI: 10.1016/S1001-0742(13)60570-7

Chapter 2

# Recent Advances in the Synthesis of Carboxylic Acid Esters

Kouichi Matsumoto, Rina Yanagi and Yohei Oe

Additional information is available at the end of the chapter

http://dx.doi.org/10.5772/intechopen.74543

**Abstract**

In this chapter, recent advances in the synthesis of carboxylic acid esters are summarized based on the utilization of carboxylic acids as electrophiles or nucleophiles in reactions. Condensation reagents or catalysts connect the carboxylic acids with the alcohols to afford the corresponding esters, together with the formation of 1 equiv. of $H_2O$, in which the carboxylic acids can be regarded as the electrophile. In contrast, the carboxylate ion intermediates derived from the carboxylic acids react with alkyl halides, carbocations, or their equivalents to produce the esters, in which the carboxylate ions from the carboxylic acids can be regarded as the nucleophile. This chapter mainly introduces the recent progress in this field of the formation of esters, based on the classification of the role of carboxylic acids in reactions.

**Keywords:** esterification, carboxylic acids, condensation reagents, catalysts, reaction media, reaction methods, $S_N2$ reactions, electrochemistry, transition metal catalysts, addition reactions

## 1. Introduction

In organic chemistry, the development of the efficient synthesis of carboxylic acid esters using carboxylic acids is still one of central research topics, because the organic material compounds, drug molecules, and natural products often contain ester unit as the functional group [1–3]. As for the view point of the synthesis of esters, the corresponding carboxylic acids are usually utilized as the key starting material and play an important role [1–3].

So far, many kinds of synthetic methods for the esters from carboxylic acids are well recognized and utilized, but a lot of researchers still have studied to investigate the new methods or aspects, because the synthesis of esters is also important from the point of view of green

chemistry and industry. In this chapter, recent advances in the synthesis of carboxylic acid esters are described. The reactions are classified into two categories, i.e., the reactions utilizing carboxylic acids as (1) electrophiles and (2) nucleophiles, in which the reactions are conducted by using various chemical reagents and catalysts as well as by using interesting reaction media and methods. Although many papers have appeared in this filed, we herein have introduced important and selected examples because of the limitation of number of pages.

## 2. Synthesis of carboxylic acid esters using carboxylic acids as electrophiles

The typical and traditional method for the synthesis of carboxylic acid esters is the reaction of carboxylic acids with an excess amount of alcohols in the presence of a catalytic amount of $H_2SO_4$ by using Dean-Stark apparatus [1–3], in which $H_2SO_4$ catalyzes the addition of the alcohol to the carboxylic acid, and the $H_2O$ thus generated is removed by Dean-Stark apparatus (**Scheme 1 (a)**). This reaction is called as Fischer esterification. However, there are some drawbacks. The excess amount of alcohols is used. The Dean-Stark apparatus is usually necessary. In addition, the substrates bearing the functional group which reacts with the acid cannot be utilized in this reaction. The alternative and reliable method to be developed is the use of DCC in the presence of a catalytic amount of DMAP (**Scheme 1 (b)**) [4]. DCC can serve as useful condensation reagents. The use of DCC as the condensation reagent realizes the decrease of the amount of alcohols. In addition, Mitsunobu reaction is also reliable method [5–8].

Besides the use of DCC, other condensation reagents are also developed. 2-Halo-pyridinium salts called Mukaiyama condensation reagent serve as effective reagents [9]. Mukaiyama et al. have extensively contributed this research area for the development of useful condensation reagents [10]. In addition, BOP ((benzotriazol-1-yloxy)-tris(dimethylamino)phosphonium hexafluorophosphate)) [11, 12], CDI (carbonyldiimidazole) [13, 14], DMT-MM (4-(4,6-dimethoxy-1,3,5-triazin-2-yl)-4-methylmorpholinium chloride) [15–17], and so on [18, 19] are well established and used in the ester formation reactions.

As described above, many condensation reagents have been developed so far. However, there are still reports for this filed. **Table 1** shows recent and selected reports of condensation reactions between carboxylic acids and alcohols using a stoichiometric amount of condensation

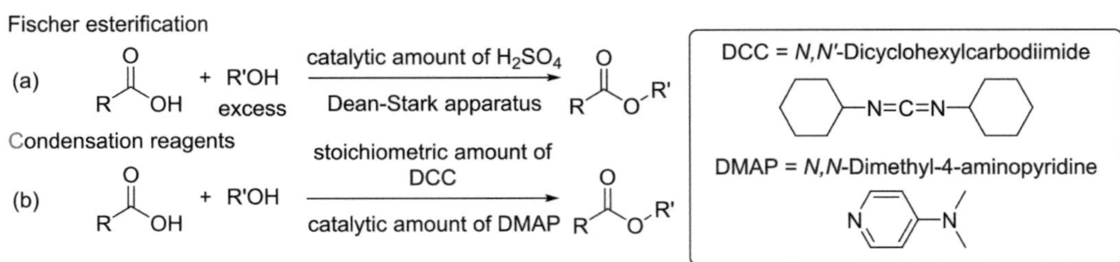


**Scheme 1.** Typical and traditional procedures for the synthesis of carboxylic acid esters and their chemical structures of reagents.

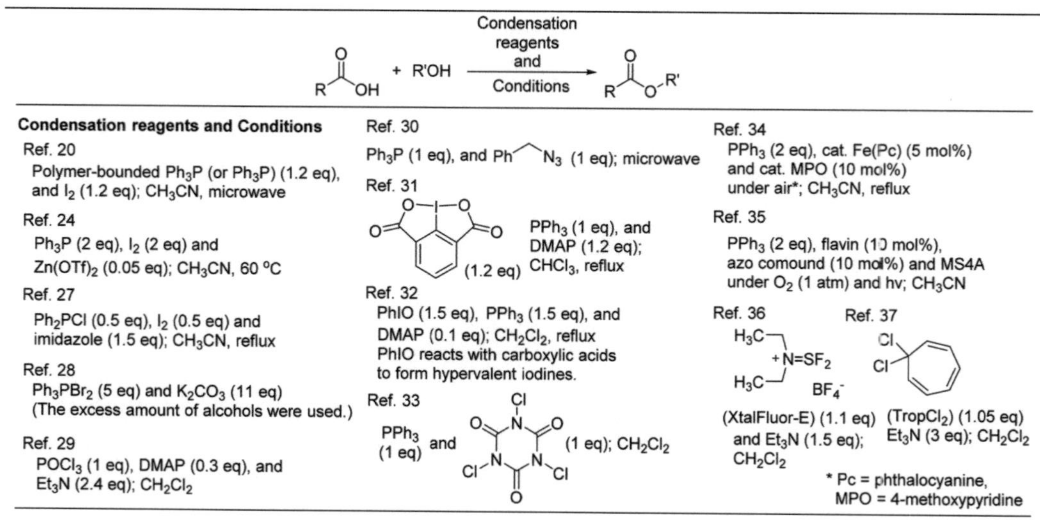


**Table 1.** Various condensation reagents and conditions used for the synthesis of carboxylic acid esters under the nearly equimolar carboxylic acids and alcohols (recent and selected examples).

reagents. Basically, the amount of alcohols is not excess (**Table 1**). For examples, polymer-bounded $Ph_3P$ (or $Ph_3P$)/$I_2$/microwave [20–23] and $Ph_3P/I_2/Zn(OTf)_2$ condition [24] have been reported, because the combination of $PPh_3$ and $I_2$ in the presence of the base has been well known so far [25, 26]. In addition, the use of $Ph_2PCl/I_2$/imidazole [27], $Ph_3PBr_2$ [28], $POCl_3$/DMAP/$Et_3N$ [29], or $PPh_3/BnN_3$/microwave (Staudinger's phosphazene) [30] systems was found to be effective for the esterification. The hypervalent iodine reagents could be utilized in coupling reactions [31, 32]. The combination of $PPh_3$ and trichloroisocyanuric acid was also effective [33]. These reactions are based on the activation of $PPh_3$. The catalytic activation of $PPh_3$ can be achieved by iron [34]. Photo-irradiated procedure in the presence of $PPh_3$ with a catalytic amount of flavin and azo compound under $O_2$ was developed [35]. XtalFluor-E and tropylium-based coupling reagents were found to be effective for the esterification [36, 37].

In the view point of green chemistry, the use of a catalytic amount of reagents is one of the attractive approaches, in which the ratio between carboxylic acids and alcohols is approximately equal. In 2000, Yamamoto et al. reported that 0.1 to 1.0 mol% hafnium (IV) salts in toluene at reflux condition catalyzed the condensation reaction of equimolar amount of carboxylic acids and alcohols (**Table 2**, entry 1) [38–47]. Since then, various types of catalysts have been found for the effective esterification reactions. Selected examples are summarized in **Table 2**. Diphenylammonium triflate (entry 2) [48], fluoroalkyldistannoxane (entry 3) [49], $HClO_4$-$SiO_2$ (entry 4) [50–52], $Ti_4^+$-mont (mont = montmorillonite, entry 5) [53], bulky diarylammonium arenesulfonates (entry 6) [54–58], $Zn(ClO_4)_2$-$6H_2O$ (entry 7) [59], pentafluorophenylammonium triflate (entry 8) [60], TsOH or CSA (entry 9) [61], phosohorofluoridic acid (entry 10) [62], *N,N*-diarylammonium pyrosulfates in $H_2O$ (entry 11) [63, 64], TfOH in Solkane365mfc (entry 12) [65], 2-oleamido-5-nitro-pyridinium *p*-toluenesulfonate (entry 13) [66], zirconocene complex (entry 14) [67], and L-leucine as an organocatalyst (entry 15) [68] have been reported for the effective catalyst for the esterification using equal or nearly equal amount of carboxylic acids and alcohols.

General reaction: RCOOH + R'OH → (Catalyst (X mol%), Solvent, Conditions) → RCOOR'

| Entry | Authors | Published year | Catalyst (X mol%) | Solvent | Conditions | Ref. |
|---|---|---|---|---|---|---|
| 1 | H. Yamamoto et al. | 2000 | hafnium(IV) salts (0.1mol%-1.0 mol%) | toluene | reflux, 5-36 h | 38 |
| 2 | Y. Tanabe et al. | 2000 | diphenylammonium triflate (DPAT) (1 mol% - 10 mol%) | toluene | 80 °C, 4-48 h | 48 |
| 3 | J. Otera et al. | 2002 | fluoroalkyldistannoxane (5 mol%) | FC-72 | 150 °C, 10-16 h | 49 |
| 4 | A. K. Chakraborti et al. | 2003 | $HClO_4$-$SiO_2$ (1 mol%) | solvent-free | 80 °C, 3.5-20 h | 50 |
| 5 | K. Kaneda et al. | 2003 | $Ti_4^+$-mont (2 mol%) | solvent-free | 110-150 °C, 3-6 h | 53 |
| 6 | K. Ishihara et al. | 2005 | bulky diarylammonium arenesulfonates (1 mol%-10 mol%) | heptane or solvent-free | r.t.-115 °C, 1-72 h | 54 |
| 7 | G. Bartoli et al. | 2005 | $Zn(ClO_4)_2$-$6H_2O$ (5 mol%) | solvent-free | 80-100 °C, 6-48 h | 59 |
| 8 | Y. Tanabe et al. | 2006 | pentafluorophenylammonium triflate (PFPAT) (1 mol%) | toluene | 80 °C, 2-48 h | 60 |
| 9 | K. Ishihara et al. | 2008 | *p*-toluenesulfonic acid (TsOH) or 10-camphorsulfonic acid (CSA) (5 mol%) | solvent-free | 60-80 °C, 3-48 h | 61 |
| 10 | T. Murai et al. | 2009 | phosphorofluoridic acid (5 mol%) | solvent-free | 100 °C, 6 h | 62 |
| 11 | K. Ishihara et al. | 2012 | *N,N*-diarylammonium pyrosulfates (5 mol%) | $H_2O$ | 60-80 °C, 6-47 h | 63 |
| 12 | N. Shibata et al. | 2013 | TfOH (0.2-1 mol%) | solkane365mfc | 80 °C, 18-48 h | 65 |
| 13 | W. Wang et al. | 2013 | 2-oleamido-5-nitro-pyridinium *p*-toluenesulfonate (1-10mol%) | isooctane | r.t.-reflux, 5-144 h | 66 |
| 14 | R. Qiu et al. | 2017 | zirconocene complex (1 mol%) | solvent-free | 80-100 °C, 12 h | 67 |
| 15 | U. Sharma et al. | 2018 | *L*-Leucine (5 mol%) | solvent-free | 80 °C, 12-24 h | 68 |

**Table 2.** Representative progresses for the synthesis of esters using equal or nearly equal amount of carboxylic acids and alcohols in the presence of the catalyst, since 2000 (selected examples).

Another approach for the esterification of carboxylic acids with alcohols (2 equiv.) was developed by Kobayashi and coworkers [69–71], in which *p*-dodecylbenzenesulfonic acid (DBSA) was used as a surfactant-type catalyst in water (**Scheme 2**). Because the micelles of DBSA are produced, the esterification reactions between carboxylic acid and alcohol proceed in the micelles. After the reaction, the micelle releases $H_2O$. The carboxylic acids and alcohols bearing the longer alkyl chains seem to be favored because of the increase of the hydrophobicity. Thus, the equilibrium between starting materials and the product lies in right side.

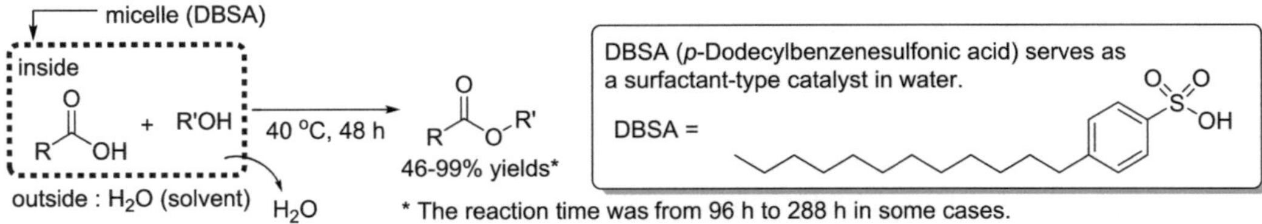


**Scheme 2.** The use of DBSA in $H_2O$ solvent (Kobayashi [69, 70]). The reactions were conducted in the ratio of RCOOH (1 equiv.) and R'OH (2 equiv.).

Porous Phenolsulphonic Acid Formaldehyde Resin (**PAFR**)

| | RCOOH | R'OH | | PAFR | RCOOR' |
|---|---|---|---|---|---|
| Condition in solvent-free: | 1 eq | 1.2-5 eq | 50 °C, 12 h | 0.70 mol% | 91-95% yields |
| Condition in $H_2O$ (solvent): | 1 eq | 5 eq | 80 °C, 48 h | 3 mol% | 80-92% yields |

**Scheme 3.** The use of PAFR as the condensation catalyst (Uozumi [72, 73]).

The use of resin was also reported by Uozumi et al., who designed and synthesized the porous phenolsulfonic acid formaldehyde resin (PAFR) from 4-hydroxybenzenesulfonic acid and formaldehyde (5 equiv.) in $H_2O$ (**Scheme 3**). The solid resin was allowed to react with RCOOH and R'OH to give the corresponding esters in good yields [72, 73]. The merit of the resin is that it can be recovered by the simple filtration and reused without the significant loss of the catalytic activity. Other type of solid catalysts bearing $SO_3H$ unit are also reported. For examples, polystyrene-supported sulfonic acid catalyst [74], SBA-15-functionalized propylsulfonic acid catalyst [75], *p*-sulonic acid calix[*n*]arenes catalyst [76], *β*-cyclodextrin-derived carbonaceous catalyst [77], and sulfonated hyperbranched poly(aryleneoxindole) acid catalyst [78] are developed and utilized for the esterification.

As for the promising reaction tool for the esterification, the use of the flow chemistry has emerged. For example, Uozumi et al. applied the PAFR to the flow method [72, 73]. Fukuyama et al. demonstrated Fisher esterification by the flow system, in which silica bearing terminal -$SO_3H$ group was used [79, 80].

## 3. Synthesis of carboxylic acid esters using carboxylic acids as nucleophiles

### 3.1. Nucleophilic reactions of carboxylate ion intermediates

#### *3.1.1. The use of bases and ionic liquids*

Because the acidity of carboxylic acids is relatively high, it is easy to generate and accumulate the carboxylate ion intermediates by the deprotonation of carboxylic acids. $S_N2$ reaction of carboxylate ions with alkyl halides is one of the most popular approaches, when carboxylate ions can be used as the nucleophiles. It was found that CsF or KF is the effective base toward carboxylic acids by Clark and Miller [81]. Since then, various reactions have been reported in these fields. Recent examples of this chemistry utilize the combination of bases (such as Huning's base, $Et_3N$ and KF) and various ionic liquids (such as imidazolium salts and phosphonium salts), summarized in **Figure 1** [82–90]. The countercations of carboxylate ions are bulky cations such as imidazolium salts and phosphonium salts, which seem to increase the reactivity of the carboxylate ions toward electrophiles.

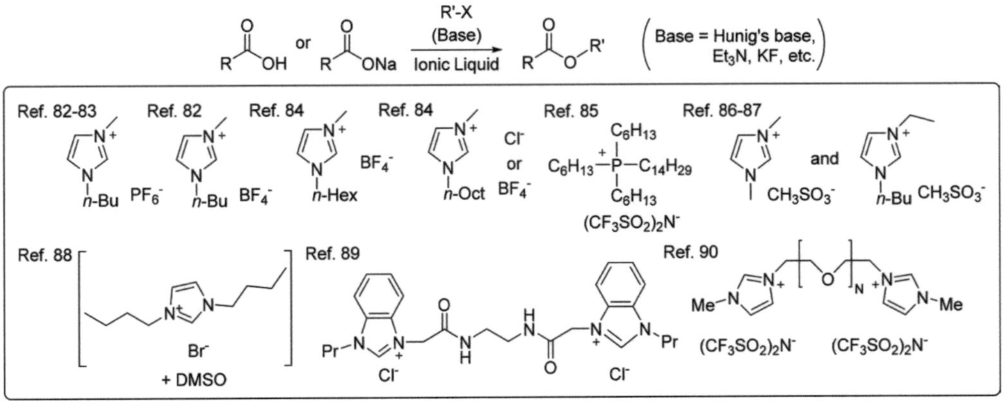

**Figure 1.** Recent and selected examples for nucleophilic reactions using carboxylate ion intermediates in various ionic liquids.

It has been well known so far that $F^-$ source such as KF and CsF can serve as good base toward carboxylic acids, described above [81]. Because of this reason, $Bu_4NF$ is also the attractive reagent for the deprotonation of carboxylic acid. In 2001, Maruoka et al. reported in situ generation of $Bu_4NF$ from the combination of a catalytic amount of $Bu_4NHSO_4$ (5 mol%) and KF-$H_2O$ (5 equiv.), which takes the proton of carboxylic acids to generate reactive carboxylate ion intermediates, whose countercation is presumably bulky $Bu_4N^+$ (**Scheme 4 (a)**) [91]. In addition, Matsumoto et al. also reported the reactions of carboxylic acids with a stoichiometric amount of $Bu_4NF$ cleanly generated and accumulated reactive carboxylate ion intermediates, which reacted with various alkyl halides to give the corresponding esters in moderate to good yields (**Scheme 4 (b)**) [92].

The use of good affinity of $F^-$ and metals such as Si and Sn was also developed in order to generate and accumulate highly reactive carboxylate ion intermediates. For example, the reaction of silyl-protected carboxylic acids with $Ph_3CF$ in the presence of a small amount of $SiF_4$ produced the corresponding esters, in which $SiF_4$ might activate both silyl-protected carboxylic acids (substrates) and $Ph_3C$-F to generate carboxylate ions and $Ph_3C^+$, respectively (**Scheme 5 (a)**, Noyori et al.) [93]. The combination of silyl-protected carboxylic acids and $Bu_4NF$ was also reported by Maruoka et al. (**Scheme 5 (b)**) [94]. Nozaki et al. reported the use of Sn in 1992. The intermediate bearing COO-Sn bond reacted with alkyl halides in the presence of CsF, as shown in **Scheme 5 (c)** [95].

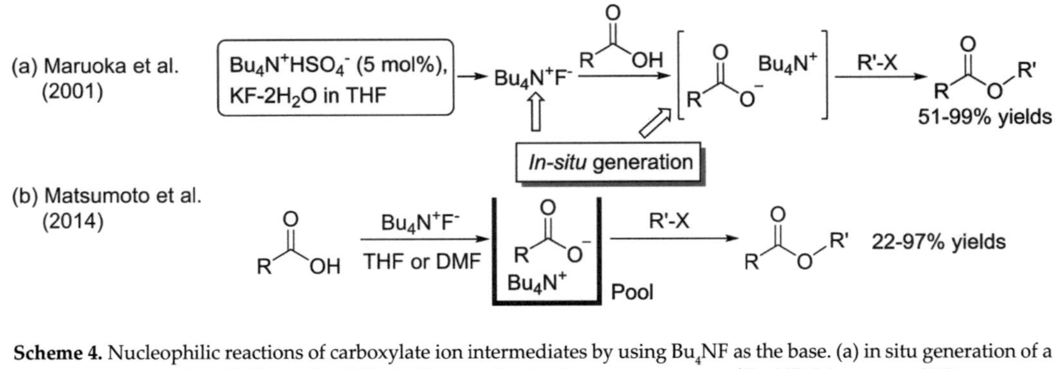

**Scheme 4.** Nucleophilic reactions of carboxylate ion intermediates by using $Bu_4NF$ as the base. (a) in situ generation of a catalytic amount of $Bu_4NF$ (Maruoka [91]). (b) the use of a stoichiometric amount of $Bu_4NF$ (Matsumoto [94]).

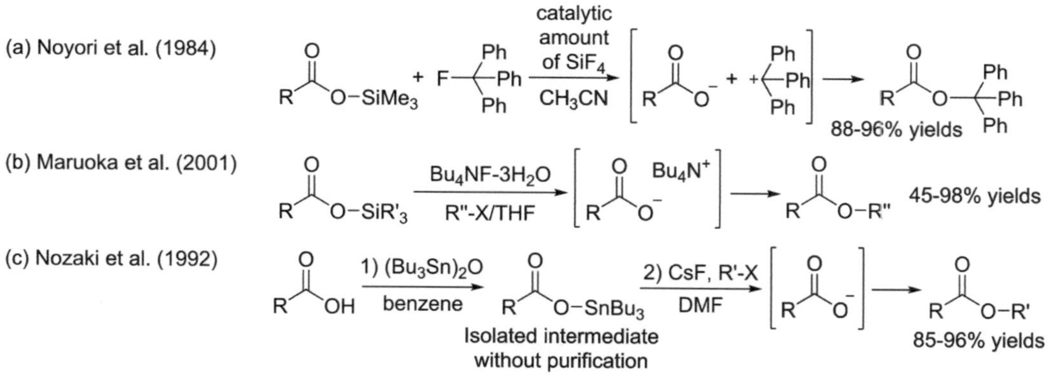

(a) Noyori et al. (1984)

(b) Maruoka et al. (2001)

(c) Nozaki et al. (1992)

Scheme 5. The use of affinity of $F^-$ and metals such as Si and Sn to generate carboxylate ion intermediates.

### 3.1.2. The use of electrochemical reduction methods

Electrochemistry is a clean technique, and basically the electron serves as the reagent instead of chemical reagents [96–99]. Therefore, electrochemistry in organic synthesis does not generate the waste derived from reagents, and is recognized as one of the powerful tools for green chemistry. The pioneering work for the esterification of carboxylic acids and alkyl halides using electrochemistry was developed by Nonaka et al. (**Scheme 6**) [100, 101]. The solution containing carboxylic acids underwent electrochemical reduction to generate highly reactive carboxylate ions, which reacted with alkyl halides to produce the corresponding esters. Matsumoto et al. investigated the detailed reaction condition, the scope and limitations, and the mechanism of the electroreductive esterification reaction (**Scheme 6**) [102].

The use of electro-generated base (EGB) [97] is also effective to generate reactive carboxylate ion intermediates, developed by Shono et al. (**Scheme 7**) [103]. 2-Pyrrolidone was electrochemically reduced and 2-pyrrolidone anion was generated and accumulated as the base in the solution phase, which reacted with carboxylic acids to generate carboxylate ions bearing the quaternary ammonium cation. The reaction was applicable to the formation of macrolides.

### 3.1.3. The use of electrophile equivalents

Some substrates were found to be effective as the electrophile equivalents, when carboxylic acids served as the nucleophile (**Figure 2**). For example, One of the interesting examples is the use of 2-benzyloxy-1-methylpyridinium triflate reported by Dudley (**Figure 2 (a)**) [104]. The benzyl cation was gradually generated, which was allowed to react with carboxylic acids. The in situ version was also established by Albiniak et al. [105]. Cu-mediated coupling reactions using aryl

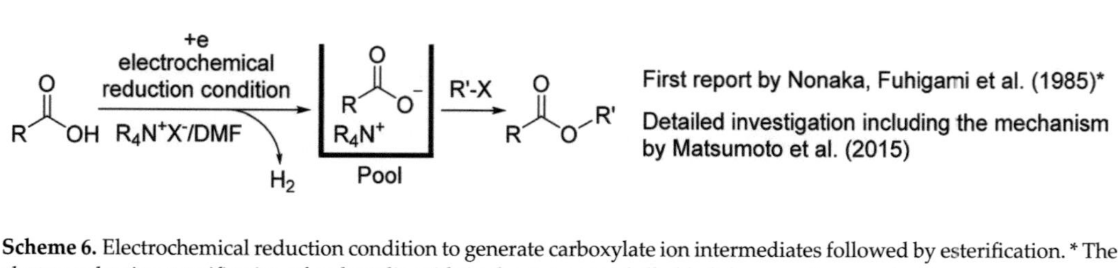

First report by Nonaka, Fuhigami et al. (1985)*

Detailed investigation including the mechanism by Matsumoto et al. (2015)

Scheme 6. Electrochemical reduction condition to generate carboxylate ion intermediates followed by esterification. * The electro-reductive esterification of carboxylic acids in the presence of alkyl halides was also examined.

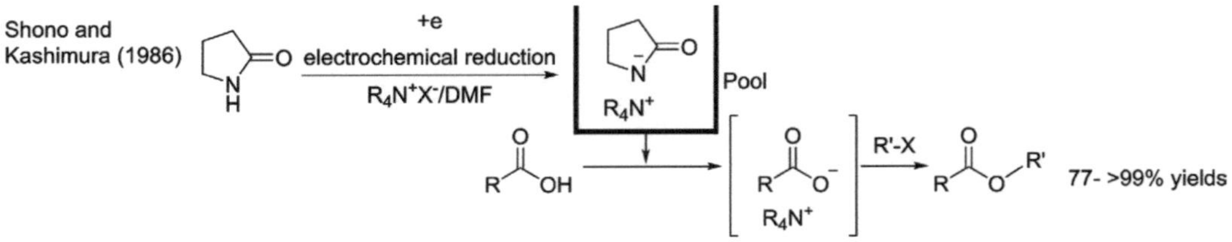

Scheme 7. Electrochemical preparation of EGB (electro-generated base) and its utilization for the esterification (Shono and Kashimura [103]).

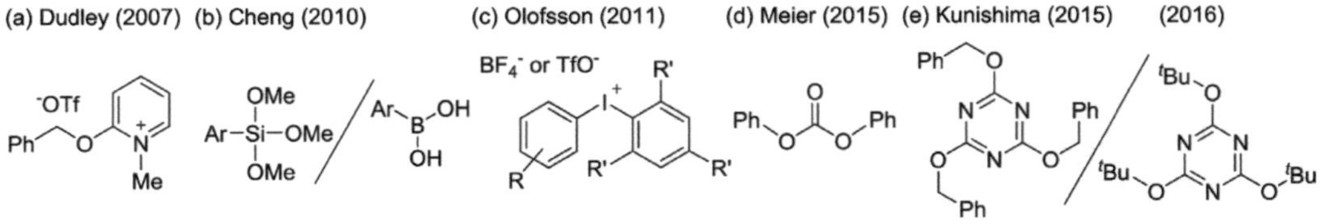

Figure 2. Selected examples for the use of electrophile equivalents.

trialkoxysilanes [106] or arylboronic acids [107] were developed by Cheng et al. (**Figure 2 (b)**). Diaryliodonium salts were found to be good electrophiles for the esterification (**Figure 2 (c)**) [108]. Meier et al. reported the use of diphenyl carbonate as the electrophile to produce phenyl esters (**Figure 2 (d)**) [109, 110]. Kunishima et al. found that 2,4,6-tris(benzyloxy)-1,3,5-triazine (TriBOT) serves as the benzyl cation equivalent via $S_N2$ mechanism in the presence of a catalytic amount of TfOH at room temperature (**Figure 2 (e)**) [111]. The reaction at high temperature also proceeded without TfOH. Thus, *O*-benzylation of carboxylic acids took place smoothly. This methodology was extended to the use of 2,4,6-tris(tert-butoxy)-1,3,5-triazine (TriAT-*t*Bu), in which carboxylic acids can react with the *tert*-butyl cation via $S_N1$ mechanism to afford the corresponding esters (**Figure 2 (e)**) [112, 113]. The reactions of the diphenylmethyl trichloroacetimidate with carboxylic acids were also investigated by Chisholm et al. [114, 115].

Recently, dimethyl sulfoxide (DMSO) was utilized for the source of $CH_3$- unit in the reaction with carboxylic acids to give the methyl esters shown in **Scheme 8**. The generation of methyl radical was indicated [116].

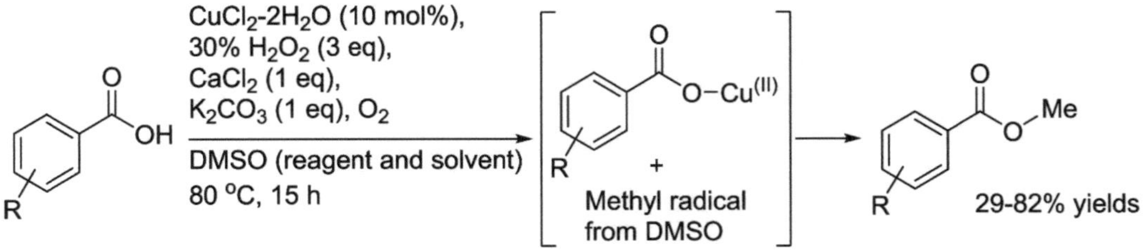

Scheme 8. Methyl esterification of carboxylic acids using DMSO (Guo [116]).

In situ generation of benzyl bromide from toluene derivatives by using $NaBrO_3/NaHSO_3$, followed by the nucleophilic reactions of the carboxylic acids could be achieved by Khan et al. (**Scheme 9**) [117]. Although aliphatic carboxylic acids were not suitable, the aromatic carboxylic acids can be converted to the corresponding esters.

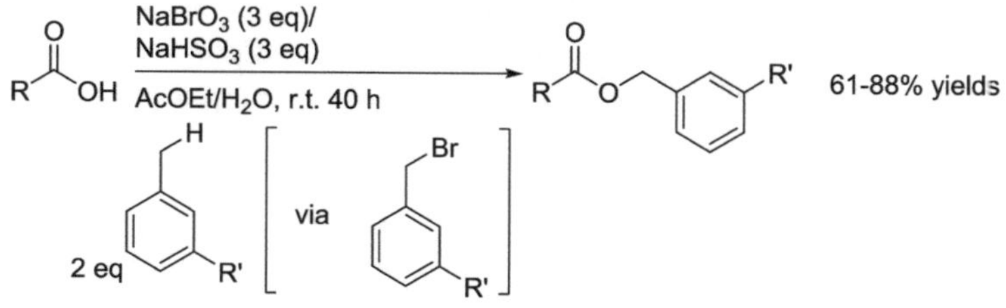

**Scheme 9.** In situ formation of benzyl bromide as the electrophile equivalent (Khan [117]).

### 3.2. The use of the functionalization of C--H bonds

Recently, the esterification of carboxylic acids and suitable substrates via the functionalization of C-H bond has been extensively studied. For example, Zhang et al. found that the reaction of carboxylic acids and toluene in the presence $Pd(OAc)_2$ (10 mol%), $CF_3SO_3H$ (10 mol%) and *N,N*-dimethylacetamide (1 equiv.) under $O_2$ (1 atm) afforded the corresponding esters via benzylic C-H bond activation (**Scheme 10**) [118]. Other interesting examples of this approach have been extensively studied [119–125].

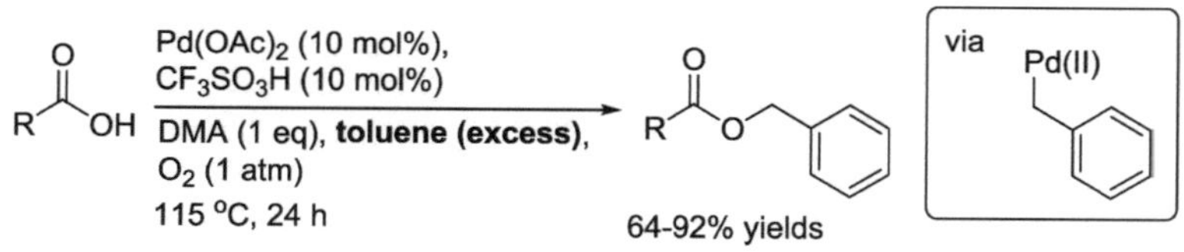

**Scheme 10.** Esterification of carboxylic acids and toluene catalyzed by $Pd(OAc)_2$ (Zhang [118]).

### 3.3. The use of metal catalysts

The addition of carboxylic acids onto C-C multiple bonds proceeds with high atom efficiency to afford the corresponding enol or alkyl esters. Hg salts have been used as the catalysts for these reactions for a long time; however, the use of these toxic salts should be avoided from the view point of green chemistry. Ruthenium complexes have been paid much attention for the alternative catalyst for the addition of carboxylic acids onto C-C multiple bonds. These ruthenium-catalyzed addition reactions of carboxylic acids to alkynes and several catalytic formation of alkyl esters by the addition of carboxylic acids to alkenes are summarized in this section.

### *3.3.1. The addition of carboxylic acids onto alkynes*

The addition reaction of carboxylic acids onto alkynes with ruthenium catalysts through the Markovnikov's rule was well investigated by Mitsudo and Dixneuf, independently [126–128]. In 1987, Mitsudo and coworkers reported that the reaction of carboxylic acids including *N*-protected $\alpha$-amino acids with alkynes were performed in the presence of Ru($\eta^5$-cod)$_2$, phosphine ligands and maleic anhydride in toluene to afford the corresponding enol esters in 31–99% yields (**Scheme 11**) [126]. This reaction proceeds with the high regioselectivity (>89% selectivity).

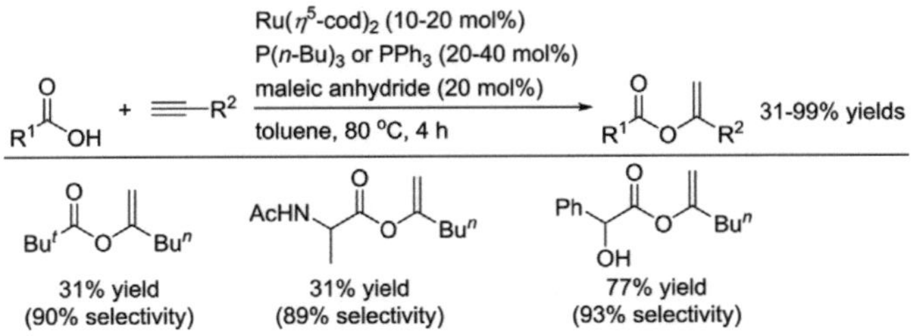

Ru($\eta^5$-cod)$_2$ (10-20 mol%)
P($n$-Bu)$_3$ or $PPh_3$ (20-40 mol%)
maleic anhydride (20 mol%)
toluene, 80 °C, 4 h
31-99% yields

31% yield (90% selectivity)
31% yield (89% selectivity)
77% yield (93% selectivity)

**Scheme 11.** Markovnikov addition (Mitsudo [126]).

On the other hand, Dixneuf and coworkers developed the simpler catalytic system for the Markovnikov addition. Thus, ($p$-cymene)$RuCl_2(PPh_3)$ catalyzed the addition of carboxylic acids to the alkynes to give the corresponding adducts in high yields (78–92%), as shown in **Scheme 12** [127, 128]. They also found the highly selective *anti*-Markovnikov addition by the use of Ru($\eta^3$-methallyl)$_2$(diphosphine) (diphosphine = dppb or dppe) to obtain the Z-enol esters in high yields with regio- and Z-selectivities (**Scheme 13**) [129–132].

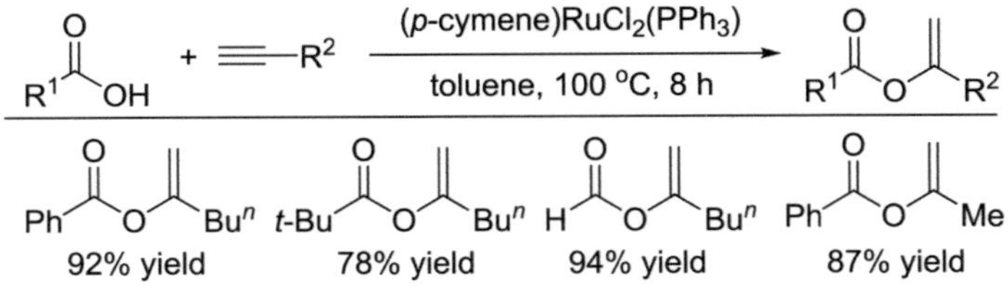

($p$-cymene)$RuCl_2(PPh_3)$
toluene, 100 °C, 8 h

92% yield
78% yield
94% yield
87% yield

**Scheme 12.** Markovnikov addition (Dixneuf [127, 128]).

The reaction mechanisms for both regioselective additions were proposed by Dixneuf et al. (**Scheme 14**), and the formation of vinylidene complex is crucial for the *anti*-Markovnikov addition. Thus, the Markovnikov adducts are formed through the activation of alkynes by the formation of $\pi$-complex with ruthenium catalyst, followed by the nucleophilic attack of the carboxylate ion onto the internal carbon atom of alkynes and the protonolysis of Ru-C $\delta$-bond [128]. On the other hand, the *anti*-Markovnikov adducts are afforded by the vinylidene complex formation between the ruthenium catalyst and alkynes, followed by the nucleophilic attack of the carboxylate ion to the terminal carbon of alkynes and the protonolysis of the resulting intermediate [129].

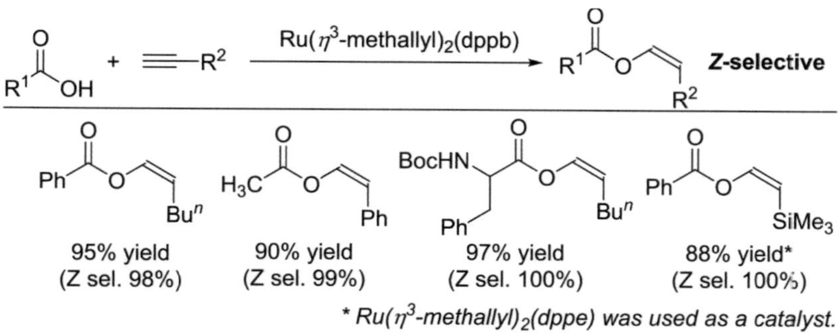

http://dx.doi.org/10.5772/intechopen.74543

**Scheme 13.** *Anti*-Markovnikov addition of carboxylic acids onto terminal alkynes (Dixneuf [129–132]).

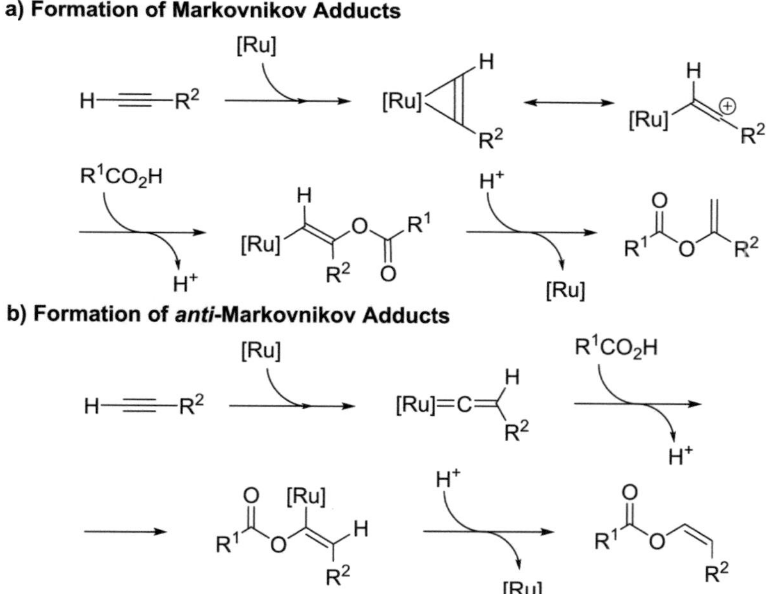


**Scheme 14.** Reaction mechanisms of the present enol ester formations.

Two simple catalytic systems, in which the regioselectivity is easily controlled by the use of the same or similar catalyst, have been reported [133, 134]. Goossen et al. found that the reaction of carboxylic acids and alkenes with $RuCl_2$(*p*-cymene)/phosphine catalyst in the presence of $K_2CO_3$ gave the corresponding Markovnikov adducts in good to excellent yields, whereas the use of DMAP instead of $K_2CO_3$ afforded the *anti*-Markovnikov (*Z*) esters in high yields (**Scheme 15**) [133]. The reactions with $RuClH(CO)(PCy)_3$ catalyst showed the interesting solvent-controlled regioselectivity (**Scheme 16**) [134]. Thus, the reaction of benzoic acid with aryl acetylene in the presence of $RuClH(CO)(PCy)_3$ as a catalyst in $CH_2Cl_2$ gave the Markovnikov adducts in high yields. On the other hand, the use of THF as a solvent instead of $CH_2Cl_2$ led to the inverse of the selectivity to afford the corresponding *anti*-Markovnikov (*Z*) esters in high

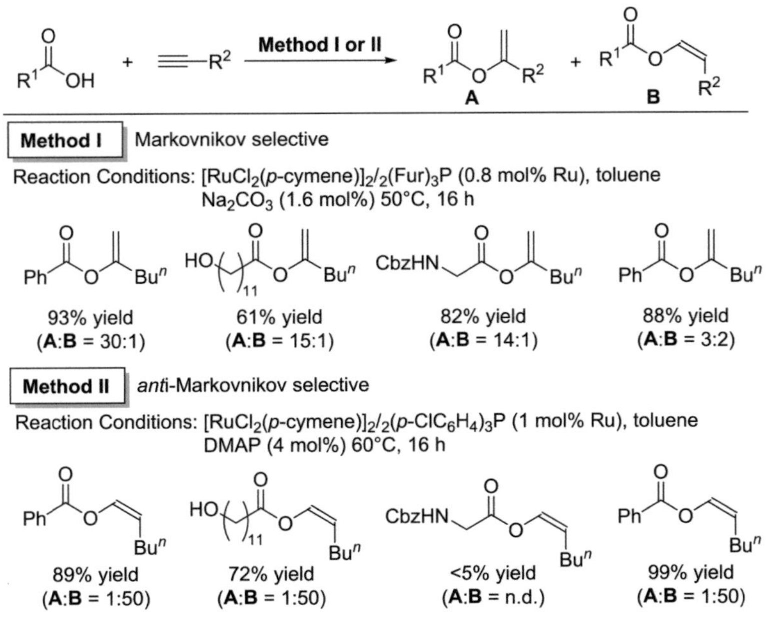

Method I or II

**Method I** Markovnikov selective

Reaction Conditions: $[RuCl_2(p\text{-cymene})]_2/2(Fur)_3P$ (0.8 mol% Ru), toluene $Na_2CO_3$ (1.6 mol%) 50°C, 16 h

93% yield (**A**:**B** = 30:1) | 61% yield (**A**:**B** = 15:1) | 82% yield (**A**:**B** = 14:1) | 88% yield (**A**:**B** = 3:2)

**Method II** *anti*-Markovnikov selective

Reaction Conditions: $[RuCl_2(p\text{-cymene})]_2/2(p\text{-ClC}_6H_4)_3P$ (1 mol% Ru), toluene DMAP (4 mol%) 60°C, 16 h

89% yield (**A**:**B** = 1:50) | 72% yield (**A**:**B** = 1:50) | <5% yield (**A**:**B** = n.d.) | 99% yield (**A**:**B** = 1:50)

**Scheme 15.** Regioselective addition (Goossen [133]).

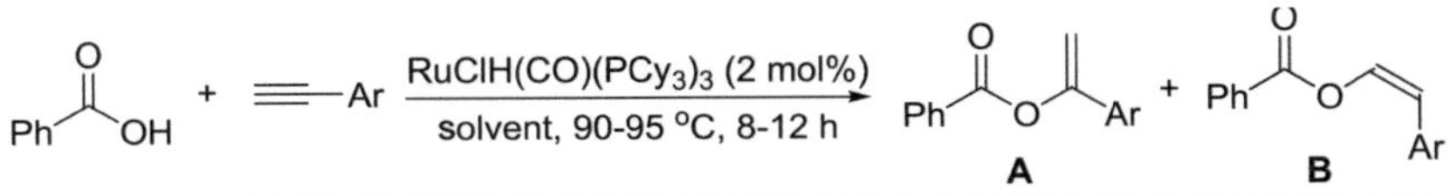

$RuClH(CO)(PCy_3)_3$ (2 mol%), solvent, 90-95 °C, 8-12 h

| Ar | Solvent | Yield of **A** | Yield of **B** |
|---|---|---|---|
| Ph | $CH_2Cl_2$ | 98 | 0 |
| | THF | 13 | 85 |
| *p*-$MeOC_6H_4$ | $CH_2Cl_2$ | 91 | 0 |
| | THF | 0 | 95 |

**Scheme 16.** Solvent-controlled selective additions (Yi [134]).

yields together with the high selectivity. It is considered that the use of THF accelerated the formation of vinylidene complex intermediates. Unfortunately, this interesting effect was not observed in the reaction of aliphatic alkynes.

The stereoselective formation of *anti*-Markovnikov (*E*) esters is so far limited [135–137]. Verpoort [135], Leong [136] and Fan [137] and their coworkers showed the (*E*)-selective formation of enol esters with their own ruthenium catalyst (**Scheme 17**), independently. However, the *E*/*Z* ratios were dependent on the alkyne and/or substrates of carboxylic acids. The rhenium catalyst showed the *E*-selectivity, though the selectivity was also strongly dependent on the substrates of alkynes. Similar selectivity was obtained when $Re(CO)_5Br$ was used as a catalyst, though the *E*/*Z* ratios were moderate (**Scheme 18**) [138].

Two examples for the regio- and *E*-selective addition of carboxylic acids onto "internal" alkynes have been reported (**Scheme 19**). Lang et al. found that the reaction of the carboxylic acid with

Ru catalyst (1.06 mol%), in toluene, 110 °C — E and Z enol esters

| Verpoort et al. (2002) | Leong et al. (2006) | Fan et al. (2010) |
|---|---|---|
| 86% yield ($E/Z$ = 4.0) | 76% yield ($E/Z$ = 6.0) | 89% yield ($E/Z$ = 5.0) |
| 1.06 mol% catalylst in toluene, 110 °C | 1 mol% catalyst in toluene, 110 °C | 1 mol% catalyst neat, 75 °C |

**Scheme 17.** Ruthenium catalysts for *E*-selective additions.

$Re(CO)_5Br$ (1 mol%), in *n*-heptane, 110 °C

| R | Yield | ratio (*E*/*Z*) |
|---|---|---|
| Ph | 72 | 27/73 |
| *n*-Bu | 78 | 67/33 |

**Scheme 18.** Rhenium catalysis for additions of carboxylic acids onto alkynes (Hua [138]).

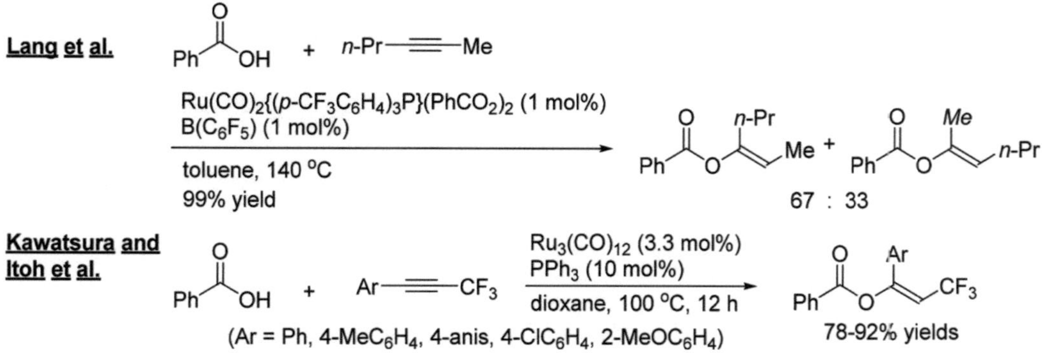


**Scheme 19.** Regio- and stereoselective additions to unsymmetrical internal alkynes.

symmetrical internal alkynes in the presence of $[Ru(CO)_2\{(p\text{-}CF_3C_6H_4)_3P\}(O_2CPh)_2]/B(C_6F_5)_3$ catalyst afforded the corresponding *E*-enol esters in up to 99% yield with the extremely high *E*-selectivity [139]. This catalytic system also achieved the regio- and stereoselective addition of carboxylic acids to unsymmetrical internal alkynes, in which the *E*/*Z* ratio reached up to 72:28. Kawatsura and Itoh reported the reaction with alkynes having trifluoromethyl and aryl group [140]. The reaction proceeded with high regioselectivity and stereoselectivity to provide the corresponding (*E*)-enol esters including trifluoromethyl group in up to 92% yield.

### *3.3.2. The addition of carboxylic acids onto alkenes*

In 2004, Oe et al. reported the first transition metal-catalyzed addition of carboxylic acids onto alkenes. Thus, the reaction of benzoic acids with norbornene was carried out in the presence of $[Cp^*RuCl_2]_2$/AgOTf/Dppb catalyst in toluene at 85 °C to obtain the corresponding norbornyl benzoates in good to high yields (**Scheme 20**) [141]. After that, several metal catalysts for the addition of carboxylic acids to alkenes have been reported and are summarized in **Table 3.**

Norbornene is a generally good substrate due to the strain of the C–C double bond, therefore the reported catalyst afforded the corresponding esters in high yields (**Table 3**) [142–145]. In 2005, He et al. reported the Au catalyst, where the four alkenes including unstrained 1-octene were transformed into the corresponding esters in 75–95% yields (entry 1) [142]. Hii et al. showed the catalytic activity of $Cu(OTf)_2$, though only norbornene was used as an alkene substrate (entry 2) [143]. In these cases, the addition of phenols onto alkenes was also catalyzed under the similar reaction conditions. With only norbornene, $In(OTf)_3$ was also found as a good catalyst under the solvent-free reaction condition (entry 3) [144]. An ubiquitous iron catalysis has been reported by Sakakura et al., where the unstable ester such as acrylates can be synthesized under the solvent-free reaction conditions (entry 4) [145]. Modified ruthenium(II) catalysis including xantphos ligand improved the scope of substrates of alkenes compared to that with ruthenium(III) catalyst to afford the corresponding esters in up to 99% yield by Oe et al. (entry 5) [146].

$[Cp^*RuCl_2]_2$ (1 mol%)
AgOTf (4 mol%)
Dppb (2 mol%)
toluene, air, 85 °C, 18 h
Ar OH
O
Ar O
63-91% yields

**Scheme 20.** Ruthenium-catalyzed additions of carboxylic acids to norbornene (Oe [141]).

$R^1COOH$ + $CH_2{=}CHR^2$ → (Catalyst (X mol%), Conditions) → $R^1COOCH(CH_3)R^2$

| Entry | Authors | Published year | Catalyst (X mol%) & Conditions | Alkenes (yield, %) | Ref. |
|---|---|---|---|---|---|
| 1 | C. He et al. | 2005 | catalyst: $Ph_3PAuOTf$ (5 mol%)<br>conditions: in toluene, 85 °C | norbornene (>95%)<br>and 4 alkenes | 142 |
| 2 | K. K. Hii et al. | 2005 | catalyst: $Cu(OTf)_2$ (5 mol%)<br>conditions: in dioxane, 80 °C | norbornene (80-98%) | 143 |
| 3 | W. Chen et al. | 2007 | catalyst: $In(OTf)_3$ (5 mol%)<br>conditions: neat, 80 °C, 2 h | norbornene (80-92%) | 144 |
| 4 | T. Sakakura et al. | 2008 | catalyst: $Fe(OTf)_3$ (2 mol%)<br>conditions: in $Bu_2O$, 80 °C, 18 h | norbornene (98-99%)<br>cyclohexene (70-88%)<br>1-octene (78%) | 145 |
| 5 | Y. Oe et al. | 2015 | catalyst: $RuCl_2(xan)\{P(OPh)_3\}_3$ (5 mol%)<br>AgOTf(10 mol%)<br>conditions: in toluene, 80 °C | 4-allylanisole (>99%)<br>and 4 alkenes | 146 |

**Table 3.** Recent reported catalysts and generality of alkene substrates.

Hartwig and He found that TfOH itself showed the catalytic activity for the present addition reactions (**Scheme 21**) [147, 148]. Interestingly, the relatively large amount of catalyst and/or higher reaction temperature decreased the chemical yield of the product. It might be due to the polymerization of substrates of alkenes. Accompanied with the importance of triflate in the metal-catalyzed reaction described above, these metal catalysts might act as a TfOH source.

Recently, an organocatalytic addition under the LED light irradiation conditions have been reported by Nicewicz et al. (**Scheme 22**) [149]. The reaction of carboxylic acids and internal and/or cyclic alkenes proceeds nicely under mild conditions to afford the corresponding esters in up to 99% yield regio-selectively.

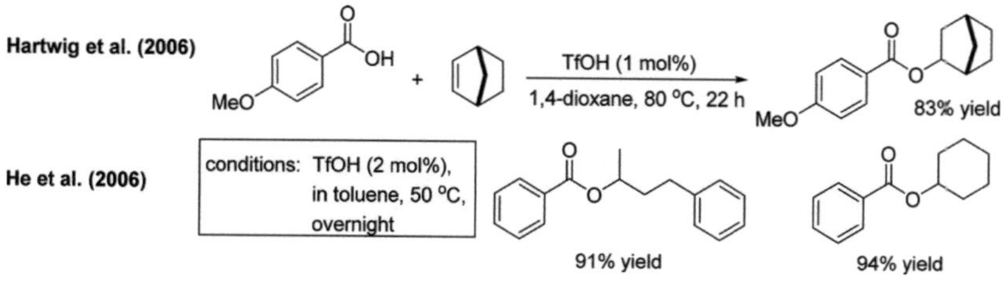

**Scheme 21.** TfOH-catalyzed reactions.

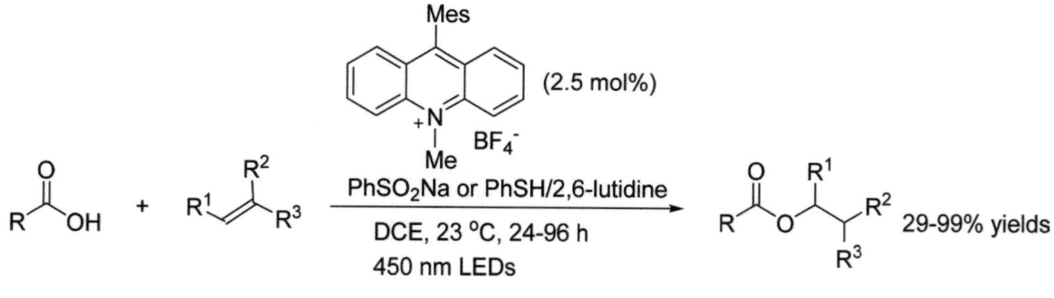

**Scheme 22.** The additions of carboxylic acids onto alkenes in the presence of the organocatalyst under the LED irradiation (Nicewicz [149]).

## 4. Conclusion

In this chapter, the recent progress of the esterification reactions using carboxylic acids as the starting material was overviewed, together with some basic and pioneering works. Various reagents, catalysts, synthetic media, and methods have been developed so far, and the quality of this field seems to be obviously increased. Because the topic of the efficient synthesis of esters is still an important task, it is expected that more innovative approach is discovered in near future.

## Author details

Kouichi Matsumoto[1*#], Rina Yanagi[1] and Yohei Oe[2]

*Address all correspondence to: kmatsumo@chem.kindai.ac.jp

1 Department of Chemistry, School of Science and Engineering, Kindai University, Higashi-osaka, Osaka, Japan

2 Department of Biomedical Information, Faculty of Life and Medical Sciences, Doshisha University, Kyotanabe, Kyoto, Japan

#This book chapter is dedicated to K. M.'s wife, Yuko Matsumoto

## References

[1] Otera J, Nishikido J. Esterification: Methods, Reactions, and Applications, 2nd ed. Weinheim, Germany: Wiley-VCH Verlag GmbH & Co. KGaA; 2009. DOI: 10.1002/9783527627622

[2] Siengalewicz P, Mulzer J, Rinner U. Synthesis of Esters and Lactones. In: Knochel P, Molander GA, editors. Comprehensive Organic Synthesis. 2nd ed; 2014;**6**:355-410. DOI: 10.1016/B978-0-08-097742-3.00612-1

[3] Li JJ, Corey EJ, editors. Carboxylic Acid Derivatives Synthesis, in Name Reactions for Functional Group Transformations. Hoboken, NJ, USA: John Wiley & Sons, Inc; 2007. DOI: 10.1002/9780470176511.ch6

[4] Neises B, Steglich W. Simple method for the esterification of carboxylic acids. Angewantde Chemie International Edition England. 1978;**17**:522-524. DOI: 10.1002/anie.197805221. DIC (*N,N'*-diisopropylcarbodiimide) is also recognized as a useful reagent for the alternation of DCC

[5] Mitsunobu O, Yamada M. Preparation of esters of carboxylic and phosphoric acid via quaternary phosphonium salts. Bulletin Chemical Society of Japan. 1967;**40**:2380-2382. DOI: 10.1246/bcsj.40.2380

[6] Hirose D, Gazvoda M, Kosmrlj J, Taniguchi T. Advances and mechanistic insight on the catalytic Mitsunobu reaction using recyclable azo reagents. Chemical Science. 2016;**7**:5148-5159. DOI: 10.1039/C6SC00308G. Recent reports of catalytic versions of Mitsunobu reaction

[7] Hirose D, Taniguchi T, Ishibashi H. Recyclable Mitsunobu reagents: Catalytic Mitsunobu reactions with an iron catalyst and atmospheric oxygen. Angewandte Chemie, International Edition. 2013;**52**:4613-4617. DOI: 10.1002/anie.201300153. Recent reports of catalytic versions of Mitsunobu reaction

[8] But TYS, Toy PH. Organocatalytic Mitsunobu reactions. Journal of the American Chemical Society. 2006;**128**:9636-9637. DOI: 10.1021/ja063141v. Recent reports of catalytic versions of Mitsunobu reaction

[9] Mukaiyama T, Usui M, Shimada E, Saigo K. Convenient method for the synthesis of carboxylic esters. Chemistry Letters. 1975:1045-1048. DOI: 10.1246/cl.1975.1045

[10] Funasaka S, Mukaiyama T. A versatile, practical, and inexpensive reagent, pyridine-3-carboxylic anhydride (3-PCA), for condensation reactions. Bulletin Chemical Society of Japan. 2008;**81**:148-159. DOI: 10.1246/bcsj.81.148, and references therein

[11] Castro B, Dormoy J, Evin G, Selve C. Reactifs de couplage peptidique I (1) - l'hexafluorophosphate de benzotriazolyl *N*-oxytrisdimethylamino phosphonium (B.O.P.). Tetrahedron Letters. 1975;**16**:1219. DOI: 10.1016/S0040-4039(00)72100-9

[12] Kim MH, Patel DV. BOP as a reagent for mild and efficient preparation of esters. Tetrahedron Letters. 1994;**31**:5603-5606. DOI: 10.1016/S0040-4039(00)77257-1

[13] Paul R, Anderson GW. *N,N*'-Carbonyldiimidazole, a new peptide-forming reagent. Journal of the American Chemical Society. 1960;**82**:4596-4600. DOI: 10.1021/ja01502a038

[14] For example of reviews, Armstrong A, Li W. N,N′-Carbonyldiimidazole. e-EROS Encyclopedia of Reagents for Organic Synthesis. 2007. DOI: 10.1002/9780470842898.rc024.pub2

[15] Kunishima M, Kawachi C, Iwasaki F, Terao K, Tani S. Synthesis and characterization of 4-(4,6-dimethoxy-1,3,5-triazin-2-yl)-4-methylmorpholinium chloride. Tetrahedron Letters. 1999;**40**:5327-5330. DOI: 10.1016/S0040-4039(99)00968-5

[16] Kunishima M, Morita J, Kawachi C, Iwasaki F, Terao K, Tani S. Esterification of carboxylic acids with alcohols by 4-(4,6-dimethoxy-1,3,5-triazin-2-yl)-4-methylmorpholinium chloride (DMTMM). Synlett. 1999:1255-1256. DOI: 10.1055/s-1999-2828

[17] Kunishima M, Kawachi C, Morita J, Terao K, Iwasaki F, Tani S. 4-(4,6-Dimethoxy-1,3,5-triazin-2-yl)-4-methylmorpholinium chloride: An efficient condensing agent leading to the formation of amides and esters. Tetrahedron. 1999;**55**:13159-13170. DOI: 10.1016/S0040-4020(99)00809-1

[18] Valeur E, Bradley M. One of the excellent reviews for condensation regents including uronium based coupling reagents. Chemical Society Reviews. 2009;**38**:606-631. DOI: 10.1039/b701677h

[19] The following catalog was referred for making this paragraph. Condensation reagents catalog for organic synthesis, 2nd ed, Wako Pure Chemical Industries (written in Japanese), and references therein

[20] Pathak G, Das D, Rokhum L. A microwave-assisted highly practical chemoselective esterification and amidation of carboxylic acids. RSC Advances. 2016;**6**:93729-93740. DOI: 10.1039/C6RA22558F

[21] Choi MKW, He HS, Toy PH. Direct radical polymerization of 4-styryldiphenylphosphine: Preparation of cross-linked and non-cross-linked Triphenylphosphine-containing polystyrene polymers. The Journal of Organic Chemistry. 2003;**68**:9831-9834. DOI: 10.1021/jo035226+. See also, the polymer-supported $PPh_3$ mediated esterification

[22] Jaita S, Phakhodee W, Pattarawarapan M. Ultrasound-assisted methyl esterification of carboxylic acids catalyzed by polymer-supported triphenylphosphine. Synlett. 2015;**26**: 2006-2008. DOI: 10.1055/s-0034-1381117. See also, the polymer-supported $PPh_3$ mediated esterification

[23] Pathak G, Rokhum L. Selective monoesterification of symmetrical diols using resin-bound triphenylphosphine. ACS Combinatorial Science. 2015;**17**:483-487. DOI: 10.1021/acscombsci.5b00086. See also, the polymer-supported $PPh_3$ mediated esterification

[24] Mamidi N, Manna D. $Zn(OTf)_2$-promoted chemoselective esterification of hydroxyl group bearing carboxylic acids. The Journal of Organic Chemistry. 2013;**78**:2386-2396. DOI: 10.1021/jo302502r

[25] Garegg PJ, Samuelsson B. Novel reagent system for converting a hydroxy group into an iodo group in carbohydrates with inversion of configuration. Journal of the Chemical Society, Chemical Communications. 1979:978-980. DOI: 10.1039/c39790000978

[26] Morcillo SP, de Cienfuegos LA, Mota AJ, Justicia J, Robles R. Mild method for the selective esterification of carboxylic acids based on the Garegg-Samuelsson reaction. The Journal of Organic Chemistry. 2011;**76**:2277-2281. DOI: 10.1021/jo102395c

[27] Nowrouzi N, Mehranpour AM, Rad JA. A simple and convenient method for preparation of carboxylic acid alkyl esters, phenolic and thioesters using chlorodiphenylphosphine/$I_2$ and imidazole reagent system. Tetrahedron. 2010;**66**:9596-9601. DOI: 10.1016/j.tet.2010.10.022

[28] Salome C, Kohn H. Triphenylphosphine dibromide: A simple one-pot esterification reagent. Tetrahedron. 2009;**65**:456-460. DOI: 10.1016/j.tet.2008.10.062

[29] Chen H, Xu X, Liu L, Tang G, Zhao Y. Phosphorus oxychloride as an efficient coupling reagent for the synthesis of esters, amides and peptides under mild conditions. RSC Advances. 2013;**3**:16247-16250. DOI: 10.1039/c3ra42887g

[30] Dinesh M, Ranganathan R, Archana S, Sathishkumar M, Roshan B, Mohamed S, Ponnuswamy A. Staudinger's phosphazene as an efficient esterifying reagent. Synthetic Communication. 2016;**46**:1454-1460. DOI: 10.1080/00397911.2016.1211705

[31] Tian J, Gao W, Zhou D, Zhang C. Recyclable hypervalent iodine(III) reagent iodosodilactone as an efficient coupling reagent for direct esterification, amidation, and peptide coupling. Organic Letters. 2012;**14**:3020-3023. DOI: 10.1021/ol301085v

[32] Boulogeorgou MA, Triantakonstanti VV, Gallos Jn K. The system PhIO/$Ph_3P$ as an efficient reagent for mild and direct coupling of alcohols with carboxylic acids. ARKIVOC 2015:244-251. DOI: 10.3998/ark.5550190.p008.997

[33] Rodrigues RC, Barros IMA, Lima ELS. Mild one-pot conversion of carboxylic acids to amides or esters with $Ph_3P$/trichloroisocyanuric acid. Tetrahedron Letters. 2005;**46**:5945-5947. DOI: 10.1016/j.tetlet.2005.06.127

[34] Taniguchi T, Hirose D, Ishibashi H. Esterification via iron-catalyzed activation of triphenylphosphine with air. ACS Catalysis. 2011;**1**:1469-1474. DOI: 10.1021/cs2003824

[35] Marz M, Chudoba J, Kohout M, Cibulka R. Photocatalytic esterification under Mitsunobu reaction conditions mediated by flavin and visible light. Organic & Biomolecular Chemistry. 2017;**15**:1970-1975. DOI: 10.1039/C6OB02770A

[36] Vandamme M, Bouchard L, Gilbert A, Keita M, Paquin J. Direct esterification of carboxylic acids with perfluorinated alcohols mediated by XtalFluor-E. Organic Letters. 2016; **18**:6468-6471. DOI: 10.1021/acs.orglett.6b03365

[37] Nguyen TV, Lyons DJM. A novel aromatic carbocation-based coupling reagent for esterification and amidation reactions. Chemical Communications. 2015;**51**:3131-3134. DOI: 10.1039/C4CC09539A

[38] Ishihara K, Ohara S, Yamamoto H. Direct condensation of carboxylic acids with alcohols catalyzed by hafnium(IV) salts. Science. 2000;**290**:1140-1142. DOI: 10.1126/science.290.5494.1140

[39] Ishihara K, Nakayama M, Ohara S, Yamamoto H. A green method for the selective esterification of primary alcohols in the presence of secondary alcohols or aromatic alcohols. Synlett. 2001, 1117-1120. DOI: 10.1055/s-2001-15156

[40] Ishihara K, Nakayama M, Ohara S, Yamamoto H. Direct ester condensation from a 1:1 mixture of carboxylic acids and alcohols catalyzed by hafnium(IV) or zirconium(IV) salts. Tetrahedron. 2002;**58**:8179-8188. DOI: 10.1016/S0040-4020(02)00966-3

[41] Nakayama M, Sato A, Ishihara K, Yamamoto H. Water-tolerant and reusable catalysts for direct ester condensation between equimolar amounts of carboxylic acids and alcohols. Advanced Synthesis and Catalysis. 2004;**346**:1275-1279. DOI: 10.1002/adsc.200404149

[42] Mantri, K, Nakamura R, Komura K, Sugi Y. Esterification of long chain aliphatic acids with long chain alcohols catalyzed by multi-valent metal salts. Chemistry Letters. 2015;**34**:1502-1503. DOI 10.1246/cl.2005.1502

[43] Mantri K, Komura K, Sugi Y. $ZrOCl_2$-$8H_2O$ catalysts for the esterification of long chain aliphatic carboxylic acids and alcohols. The enhancement of catalytic performance by supporting on ordered mesoporous silica. Green Chemistry. 2005;**7**:677-682. DOI: 10.1039/b504369g

[44] Chen C, Munot YS. Direct atom-efficient esterification between carboxylic acids and alcohols catalyzed by amphoteric, water-tolerant $TiO(acac)_2$. The Journal of Organic Chemistry 2005;**70**:8625-8627. DOI: 10.1021/jo051337s

[45] Sato A, Nakamura Y, Maki T, Ishihara K, Yamamoto H. Zr(IV)-Fe(III), –Ga(III), and -Sn(IV) binary metal complexes as synergistic and reusable esterification catalysts. Advanced Synthesis and Catalysis. 2005;**347**:1337-1340. DOI: 10.1002/adsc.200505083

[46] Mantri K, Komura K, Sugi Y. Efficient esterification of long chain aliphatic carboxylic acids with alcohols over $ZrOCl_2 \cdot 8H_2O$ catalyst. Synthesis. 2005:1939-1944. DOI: 10.1055/s-2005-869951

[47] Nakamura Y, Maki T, Wang X, Ishihara K, Yamamoto H. Iron(III)-zirconium(IV) combined salt immobilized on *N*-(polystyrylbutyl) pyridinium triflylimide as a reusable catalyst for a dehydrative esterification reaction. Advanced Synthesis and Catalysis. 2006;**348**:1505-1510. DOI: 10.1002/adsc.200606126

[48] Wakasugi K, Misaki T, Yamada K, Tanabe Y. Diphenylammonium triflate (DPAT): Efficient catalyst for esterification of carboxylic acids and for transesterification of carboxylic esters with nearly equimolar amounts of alcohols. Tetrahedron Letters. 2000;**41**: 5249-5252. DOI: 10.1016/S0040-4039(00)00821-2

[49] Xiang J, Orita A, Otera J. Fluorous biphasic esterification directed towards ultimate atom efficiency. Angewandte Chemie, International Edition. 2002;**41**:4117-4119. DOI: 10.1002/1521-3773(20021104)41:21<4117::AID-ANIE4117>3.0.CO;2-4

[50] Chakraborti AK, Gulhane R. Perchloric acid adsorbed on silica gel as a new, highly efficient, and versatile catalyst for acetylation of phenols, thiols, alcohols, and amines. Chemicall Communication. 2003;1896-1897. DOI: 10.1039/B304178F

[51] Chakraborti AK, Singh B, Chankeshwara SV, Patel AR. Protic acid immobilized on solid support as an extremely efficient recyclable catalyst system for a direct and atom economical esterification of carboxylic acids with alcohols. Journalnal of Organic Chemistry. 2009;**74**:5967-5974. DOI: 10.1021/jo900614s

[52] Chakraborti AK, Chankeshwara SV, Singh B. Indian patent. 2764/DEL/2007 (Dec 28, 2007)

[53] Kawabata T, Mizugaki T, Ebitani K, Kaneda K. Highly efficient esterification of carboxylic acids with alcohols by montmorillonite-enwrapped titanium as a heterogeneous acid catalyst. Tetrahedron Letters. 2003;**44**:9205-9208. DOI: 10.1016/j.tetlet.2003.10.024

[54] Ishihara K, Nakagawa S, Sakakura A. Bulky diarylammonium arenesulfonates as selective esterification catalysts. Journal of American Chemical Society. 2005;**127**:4168-4169. DOI: 10.1021/ja050223v

[55] Sakakura A, Nakagawa S, Ishihara K. Bulky diarylammonium arenesulfonates as mild and extremely active dehydrative ester condensation catalysts. Tetrahedron. 2006;**62**:422-433. DOI: 10.1016/j.tet.2005.09.059. See also

[56] Sakakura A, Watanabe H, Nakagawa S, Ishihara K. Unusual rate acceleration in Brønsted acid catalyzed dehydration reactions: Local hydrophobic environment in aggregated *N*-(2,6-diphenylphenyl)-*N*-mesitylammonium pentafluorobenzenesulfonates. Chemistry, an Asian Journal. 2007;**2**:477-483. DOI: 10.1002/asia.200600380

[57] Sakakura A, Nakagawa S, Ishihara K. Direct ester condensation catalyzed by bulky diarylammonium pentafluorobenzenesulfonates. Nature Protocools. 2007;**2**:1746-1751. DOI: 10.1038/nprot.2007.254

[58] Ishihara K. Dehydrative condensation catalyses. Tetrahedron. 2009;**65**:1085-1109. DOI: 10.1016/j.tet.2008.11.004. Review of esterification

[59] Bartoli G, Boeglin J, Bosco M, Locatelli M, Massaccesi M, Melchiorre P, Sambri L. Highly efficient solvent-free condensation of carboxylic acids with alcohols catalyzed by zinc perchlorate hexahydrate, $Zn(ClO_4)_2$-$6H_2O$. Advanced Synthesis and Catalysis. 2005;**347**:33-38. DOI: 10.1002/adsc.200404171

[60] Funatomi T, Wakasugi K, Misaki T, Tanabe Y. Pentafluorophenylammonium triflate (PFPAT): An efficient, practical, and cost-effective catalyst for esterification, thioesterification, transesterification, and macrolactone formation. Green Chemistry. 2006;**8**:1022-1027. DOI: 10.1039/b609181b

[61] Sakakura A, Koshikari Y, Ishihara K. Open-air and solvent-free ester condensation catalyzed by sulfonic acids. Tetrahedron Letters. 2008;**49**:5017-5020. DOI: 10.1016/j.tetlet.2008.06.058

[62] Murai T, Tanaka N, Higashijima S, Miura H. Phosphorofluoridic acid-catalyzed condensation reaction of carboxylic acids with 1-arylalkyl alcohols leading to esters under solvent-free conditions. Chemistry Letters. 2009;**38**:668-669. DOI: 10.1246/cl.2009.668

[63] Sakakura A, Koshikari Y, Akakura M, Ishihara K. Hydrophobic *N,N*-diarylammonium pyrosulfates as dehydrative condensation catalysts under aqueous conditions. Organic Letters. 2012;**14**:30-33. DOI: 10.1021/ol2027366

[64] Igarashi T, Yagyu D, Naitoa T, Okumura Y, Nakajo T, Mori Y, Kobayashi S. Dehydrative esterification of carboxylic acids with alcohols catalyzed by diarylammonium *p*-dodecylbenzenesulfonates in water. Applied Catalysis B-Environment. 2012;**119-120**:304-307. DOI: 10.1016/j.apcatb.2012.03.001

[65] Xu X, Azuma A, Taniguchi M, Tokunaga E, Shibata N. Efficient direct ester condensation between equimolar amounts of carboxylic acids and alcohols catalyzed by trifluoromethanesulfonic acid (TfOH) in Solkane365mfc. RSC Advances. 2013;**3**:3848-3852. DOI: 10.1039/c3ra00132f

[66] Wang W, Liu H, Xu S, Gao Y. Esterification catalysis by pyridinium *p*-toluenesulfonate revisited-modification with a lipid chain for improved activities and selectivities. Synthetic Communication. 2013;**43**:2906-2912. DOI: 10.1080/00397911.2012.749990

[67] Tang Z, Jiang Q, Peng L, Xu X, Li J, Qiu R, Au C. Zirconocene-catalyzed direct (trans) esterification of acyl acids (esters) and alcohols in a strict 1 : 1 ratio under solvent-free conditions. Green Chemistry. 2017;**19**:5396-5402. DOI: 10.1039/C7GC02174G

[68] Kumar M, Thakur K, Sharma S, Nayal OS, Kumar N, Singh B, Sharma U. Solvent-free, L-leucine-catalyzed direct dehydrative esterification of carboxylic acids with alcohols: Direct synthesis of 3-alkoxy 1(3 H)-isobenzofuranone. Asian Journal of Organic Chemistry. 2018;**7**:227-231. DOI: 10.1002/ajoc.201700523

[69] Manabe K, Sun X, Kobayashi S. Dehydration reactions in water. Surfactant-type Brønsted acid-catalyzed direct esterification of carboxylic acids with alcohols in an emulsion system. Journal of the American Chemical Society. 2001;**123**:10101-10102. DOI: 10.1021/ja016338q

[70] Manabe K, Iimura S, Sun X, Kobayashi S. Dehydration reactions in water. Brønsted acid-surfactant-combined catalyst for ester, ether, thioether, and dithioacetal formation in water. Journal of American Chemical Society. 2002;**124**:11971-11978. DOI: 10.1021/ja026241j

[71] Jing L, Li X J, Han Y C, Chu Y. The esterification in cyclohexane/DBSA/water microemulsion system. Colloids and Surface A. Physicochemical and Engineering Aspects. 2008;**326**:37-41. DOI: 10.1016/j.colsurfa.2008.05.008

[72] Baek H, Minakawa M, Yamada YMA, Han JW, Uozumi Y. In-water and neat batch and continuous-flow direct esterification and transesterification by a porous polymeric acid catalyst. Scientific Reports. 2016;**6**:25925. DOI: 10.1038/srep25925

[73] Minakawa M, Baek H, Yamada YMA, Han JW, Uozumi Y. Direct dehydrative esterification of alcohols and carboxylic acids with a macroporous polymeric acid catalyst. Organic Letters. 2013;**15**:5798-5801. DOI: 10.1021/ol4028495

[74] Manabe K, Kobayashi S. Dehydrative esterification of carboxylic acids with alcohols catalyzed by polymer-supported sulfonic acids in water. Advanced Synthesis and Catalysis. 2002;**344**:270-273. DOI: 10.1002/1615-4169(200206)344:3/4<270::AID-ADSC270>3.0.CO;2-7

[75] Karimi B, Vafaeezadeh M. SBA-15-functionalized sulfonic acid confined acidic ionic liquid: A powerful and water-tolerant catalyst for solvent-free esterifications. Chemical Communications. 2012;**48**:3327-3329. DOI: 10.1039/c2cc17702a

[76] Fernandes SA, Natalino R, Gazolla PAR, da Silva MJ, Jham GN. *p*-sulfonic acid calix[*n*]arenes as homogeneous and recyclable organocatalysts for esterification reactions. Tetrahedron Letters. 2012;**53**:1630-1633. DOI: 10.1016/j.tetlet.2012.01.078

[77] Thombal RS, Jadhav AR, Jadhav VH. Biomass derived $\beta$-cyclodextrin-$SO_3H$ as a solid acid catalyst for esterification of carboxylic acids with alcohols. RSC Advances. 2015;**5**:12981-12986. DOI: 10.1039/C4RA16699J

[78] Verheyen T, Smet M, De Borggraeve WM. Water tolerant and reusable sulfonated hyperbranched poly(aryleneoxindole) acid catalyst for solvent-free esterification. Chemistry Select. 2017;**2**:9822-9828. DOI: 10.1002/slct.201701636

[79] Furuta A, Fukuyama T, Ryu I. Efficient flow Fischer esterification of carboxylic acids with alcohols using sulfonic acid-functionalized silica as supported catalyst. Bulletin Chemical Society of Japan. 2017;**90**:607-612. DOI: 10.1246/bcsj.20170025

[80] Audubert C, Lebel H. Mild esterification of carboxylic acids via continuous flow diazotization of amines. Organic Letters. 2017;**19**:4407-4410. DOI: 10.1021/acs.orglett.7b02231. See also another example using micro reactor. This reaction is categorized into 3.1 Nucleophilic reaction of carboxylate ion intermediate

[81] Clark JH, Miller JM. Hydrogen bonding in organic synthesis. V. Potassium fluoride in carboxylic acids as an alternative to crown ether with acid salts in the preparation of phenacyl esters. Tetrahedron Letters. 1977;**18**:599. DOI: 10.1016/S0040-4039(01)92703-0. See also, the introduction of ref. 56

[82] Cardellini F, Brinchi L, Germani R, Tiecco M. Convenient esterification of carboxylic acids by $S_N2$ reaction promoted by a Protic ionic-liquid system formed in situ in solvent-free conditions. Synthetic Communications. 2014;**44**:3248-3256. DOI: 10.1080/00397911.2014.933353

[83] Gok Y, Alici B, Cetinkaya E, Ozdemir I, Ozeroglu O. Ionic liquids as solvent for efficient esterification of carboxylic acids with alkyl halides. Turkish Journal of Chemistry. 2010;**34**:187-191. DOI: 10.3906/kim-0904-39

[84] Judeh ZMA, Shen H, Chi BC, Feng L, Selvasothi S. A facile and efficient nucleophilic displacement reaction at room temperature in ionic liquids. Tetrahedron Letters. 2002;**43**:9381-9384. DOI: 10.1016/S0040-4039(02)02327-4

[85] McNulty J, Cheekoori S, Nair J, Larichev V, Capretta A, Robertson A. A mild esterification process in phosphonium salt ionic liquid. Tetrahedron Letters. 2005;**46**:3641-3644. DOI: 10.1016/j.tetlet.2005.03.169

[86] Brinchi L, Germani R, Savelli G. Ionic liquids as reaction media for esterification of carboxylate sodium salts with alkyl halides. Tetrahedron Letters. 2003;**44**:2027-2029. DOI: 10.1016/S0040-4039(03)00179-5

[87] Brinchi L, Germani R, Savelli G. Efficient esterification of carboxylic acids with alkyl halides catalyzed by fluoride ions in ionic liquids. Tetrahedron Letters. 2003;**44**:6583-6585. DOI: 10.1016/S0040-4039(03)01693-9

[88] Dighe S, Bhattad R, Kulkarni R, Jain K, Srinivasan K. A rapid and facile esterification of Na-carboxylates with alkyl halides promoted by the synergy of the combined use of DMSO and an ionic liquid under ambient conditions. Synthetic Communications. 2010;**40**:3522-3527. DOI: 10.1080/00397910903457357

[89] Muskawar PN, Thenmozhi K, Gajbhiye M, Bhagat PR. Facile esterification of carboxylic acid using amide functionalized benzimidazolium dicationic ionic liquids. Applied Catalysis A. 2014;**482**:214-220. DOI: 10.1016/j.apcata.2014.06.004

[90] Jadhav A H, Lee K, Koo S, Seo J G. Esterification of carboxylic acids with alkyl halides using imidazolium based dicationic ionic liquids containing bis-trifluoromethane sulfonimide anions at room temperature. RSC Advances. 2015;**5**:26197-26208. DOI: 10.1039/C5RA00802F

[91] Ooi T, Sugimoto H, Doda K, Maruoka K. Esterification of carboxylic acids catalyzed by in situ generated tetraalkylammonium fluorides. Tetrahedron Letters. 2001;**42**:9245-9248. DOI: 10.1016/S0040-4039(01)02035-4

[92] Matsumoto K, Shimazaki H, Miyamoto Y, Shimada K, Haga F, Yamada Y, Miyazawa H, Nishiwaki K, Kashimura S. Simple and convenient synthesis of esters from carboxylic acids and alkyl halides using tetrabutylammonium fluoride. Journal of Oleo Science. 2014;**63**:539-544. DOI: 10.5650/jos.ess13199

[93] Hashimoto S, Hayashi M, Noyori R. An easy preparation of triphenylmethyl carboxylates. Bulletin of the Chemical Society of Japan. 1984;**57**:1431-1432. DOI: 10.1246/bcsj.57.1431

[94] Ooi T, Sugimoto H, Maruoka K. Facile conversion of trialkylsilyl esters to alkyl esters mediated by tetrabutylammonium fluoride trihydrate. Heterocycles. 2001;**54**:593-596

[95] Sato T, Otera J, Nozaki H. Cesium fluoride-promoted esterification of carboxylic acids. A practical alternative to the diazomethane method and direct conversion of organotin carboxylates. The Journal of Organic Chemistry. 1992;**57**:2166-2169. DOI: 10.1021/jo00033a048

[96] Selected reviews of electro-organic chemistry, Yoshida J, Kataoka K, Horcajada R, Nagaki A. Modern strategies in electroorganic synthesis. Chemical Reviews. 2008;**108**:2265-2299. DOI: 10.1021/cr0680843. chemrev.7b00475, and references therein

[97] Selected reviews of electro-organic chemistry, Frontana-Uribe BA, Little RD, Ibanez JG, Palma A, Vasquez-Medrano R. Organic electrosynthesis: A promising green methodology in organic chemistry. Green Chemistry. 2010;**12**:2099-2119. DOI: 10.1039/c0gc00382d. chemrev.7b00475, and references therein

[98] Selected reviews of electro-organic chemistry, Sperry JB, Wright DL. The application of cathodic reductions and anodic oxidations in the synthesis of complex molecules. Chemical Society Reviews. 2006;**35**:605-621. DOI: 10.1039/b512308a. chemrev.7b00475, and references therein

[99] Selected reviews of electro-organic chemistry, Yoshida J, Shimizu A, Hayashi R. Electrogenerated cationic reactive intermediates: The pool method and further advances. Chemical Review. in press. DOI: 10.1021/acs.chemrev.7b00475, and references therein

[100] Awata T, Baizer M, Nonaka T, Fuchigami T. Cathodic esterification of carboxylic acids. Chemical Letters. 1985;**3**:371-374. DOI: 10.1246/cl.1985.371

[101] Fuchigami T, Awata T, Nonaka T, Baizer MM. Esterification, etherification, and aldol condensation using cathodically generated organic olate anions. Bulletin Chemical Society Japan. 1986;**59**:2873-2879. DOI: 10.1246/bcsj.59.2873

[102] Miyamoto Y, Yamada Y, Shimazaki H, Shimada K, Nokami T, Nishiwaki K, Kashimura S, Matsumotoa K. Esterification of carboxylic acids with alkyl halides using electroreduction Electrochemistry 2015;**83**:161-164. DOI: 10.5796/electrochemistry.83.161

[103] Shono T, Ishige O, Uyama H, Kashimura S. Electroorganic chemistry. 91. A novel base useful for synthesis of esters and macrolides. The Journal of Organic Chemistry 1986; **51**:546-549. DOI: 10.1021/jo00354a030

[104] Tummatorn J, Albiniak PA, Dudley GB. Synthesis of benzyl esters using 2-benzyloxy-1-methylpyridinium triflate. The Journal of Organic Chemistry. 2007;**72**:8962-8964. DOI: 10.1021/jo7018625

[105] Strayer TA, Culy CC, Bunner MH, Frank AR, Albiniak PA. In situ synthesis of 2-allyloxy-1-methylpyridinium triflate for the allylation of carboxylic acids. Tetrahedron Letters. 2015;**56**:6807-6809. DOI: 10.1016/j.tetlet.2015.10.085

[106] Luo F, Pan C, Qian P, Cheng J. Copper(II)-catalyzed esterification of arenecarboxylic acids with aryl- and vinyl-substituted trimethoxysilanes. Synthesis. 2010:2005-2010. DOI: 10.1055/s-0029-1218759

[107] Zhang L, Zhang G, Zhang M, Cheng J. Cu(OTf)$_2$-mediated Chan-lam reaction of carboxylic acids to access phenolic esters. The Journal of Organic Chemistry 2010;**75**:7472-7474. DOI: 10.1021/jo101558s

[108] Petersen TB, Khan R, Olofsson B. Metal-free synthesis of aryl esters from carboxylic acids and diaryliodonium salts. Organic Letters. 2011;**13**:3462-3465. DOI: 10.1021/ol2012082

[109] Kreye O, Meier MAR. Base catalyzed sustainable synthesis of phenyl esters from carboxylic acids using diphenyl carbonate. RSC advances. 2015;**5**:53155-53160. DOI: 10.1039/C5RA10206E

[110] Shieh W, Dell S, Repic O. Tetrahedron Letters 2002;**43**:5607-5609. DOI: 10.1016/S0040-4039(02)01116-4

[111] Yamada K, Yoshida S, Fujita H, Kitamura M, Kunishima M. *O*-benzylation of carboxylic acids using 2,4,6-Tris(benzyloxy)-1,3,5-triazine (TriBOT) under acidic or thermal conditions. European Journal of Organic Chemistry. 2015:7997-8002. DOI: 10.1002/ejoc.201501172

[112] Kunishima M, Fujita H, Yamada K. PCT Int. Appl. WO 2013 073681 A1, 2013

[113] Yamada K, Hayakawa N, Fujita H, Kitamura M, Kunishima M. Development of a triazine-based *tert*-Butylating reagent, TriAT-*t*Bu. European Journal of Organic Chemistry. 2016:4093-4098. DOI: 10.1002/ejoc.201600663

[114] Adhikari AA, Shah JP, Howard KT, Russo CM, Wallach DR, Linaburg MR, Chisholm JD. Convenient formation of diphenylmethyl esters using diphenylmethyl trichloroacetimidate. Synlett. 2014;**25**:283-287. DOI: 10.1055/s-0033-1340293

[115] Shah JP, Russo CM, Howard KT, Chisholm JD. Spontaneous formation of PMB esters using 4-methoxybenzyl-2,2,2-trichloroacetimidate. Tetrahedron Letters. 2014;**55**:1740-1742. DOI: 10.1016/j.tetlet.2014.01.097

[116] Jia J, Jiang Q, Zhao A, Xu B, Liu Q, Luo W, Guo C. Copper-catalyzed O-methylation of carboxylic acids using DMSO as a methyl source. Synthesis. 2016;**48**:421-428. DOI: 10.1055/s-0035-1560967

[117] Khan KM, Maharvi GM, Hayat S, Zia-Ullah CMI, Rahman A-U. An expedient esterification of aromatic carboxylic acids using sodium bromate and sodium hydrogen sulfite. Tetrahedron. 2003;**59**:5549-5554. DOI: 10.1016/S0040-4020(03)00812-3

[118] Liu H, Shi G, Pan S, Jiang Y, Zhang Y. Palladium-catalyzed benzylation of carboxylic acids with toluene via benzylic C-H activation. Organic Letters. 2013;**15**:4098-4101. DOI: 10.1021/ol401687f

[119] Selected recent examples of this chemistry. Lu B, Zhu F, Sun H, Shen Q. Esterification of the primary benzylic C-H bonds with carboxylic acids catalyzed by ionic iron(III) complexes containing an imidazolinium cation. Organic Letters. 2017;**19**:1132-1135. DOI: 10.1021/acs.orglett.7b00148/C7RA02788E, and references therein

[120] Selected recent examples of this chemistry. Zheng Y, Mao J, Rong G, Xu X. A NCS mediated oxidative C-H bond functionalization: Direct esterification between a C(sp$^3$)-H bond and carboxylic acids. Chemical Communications. 2015;**51**;8837-8840. DOI: 10.1039/C5CC02424B/C7RA02788E, and references therein

[121] Selected recent examples of this chemistry. Ren T, Xu B, Mahmood S, Sun M, Zhang S. Cobalt-catalyzed oxidative esterification of allylic/benzylic C(sp$_3$)-H bonds. Tetrahedron. 2017;**73**:2943-2948. DOI: 10.1016/j.tet.2017.04.002/C7RA02788E, and references therein

[122] Selected recent examples of this chemistry. Lu B, Zhu F, Wang D, Sun H, Shen Q. Iron-catalyzed esterification of allylic sp$^3$ C-H bonds with carboxylic acids: Facile access to allylic esters. Tetrahedron Letters. 2017;**58**:2490-2494. DOI: 10.1016/j.tetlet.2017.05.039/C7RA02788E, and references therein

[123] Selected recent examples of this chemistry. Dick AR, Hull KL, Sanford MS. A highly selective catalytic method for the oxidative functionalization of C-H bonds. Journal of the American Chemical Society. 2004;**126**:2300-2301. DOI: 10.1021/ja031543m/C7RA02788E, and references therein

[124] Selected recent examples of this chemistry. Chen X, Hao X, Goodhue C E, Yu J. Cu(II)-catalyzed functionalizations of aryl C-H bonds using $O_2$ as an oxidant. Journal of the American Chemical Society. 2006;**128**:6790-6791. DOI: 10.1021/ja061715q/C7RA02788E, and references therein

[125] Selected recent examples of this chemistry. Mou F, Sun Y, Jin W, Zhang Y, Wang B, Liu Z, Guo L, Huang J, Liu C. Reusable ionic liquid-catalyzed oxidative esterification of carboxylic acids with benzylic hydrocarbons via benzylic Csp3-H bond activation under metal-free conditions. RSC Advances. 2017;**7**:23041-23045. DOI: 10.1039/C7RA02788E, and references therein

[126] Mitsudo T, Hori Y, Yamakawa Y, Watanabe Y. Ruthenium-catalyzed selective addition of carboxylic acids to alkynes. A novel synthesis of enol esters. The Journal of Organic Chemistry. 1987;**52**:2230-2239. DOI: 10.1021/jo00387a024

[127] Ruppin C, Dixneuf HP. Synthesis of enol esters from terminal alkynes catalyzed by ruthenium complexes. Tetrahedron Letters. 1986;**52**:6323-6324. DOI: 10.1016/S0040-4039(00)87798-9

[128] Bruneau C, Neveux M, Kabouche Z, Ruppin C, Dixneuf HP. Ruthenium-catalysed additions to alkynes: Synthesis of activated esters and their use in acylation reactions. Synlett. 1991:755-763. DOI: 10.1055/s-1991-20866

[129] Bruneau C, Dixneuf HP. Metal vinylidenes and allenylidenes in catalysis: Applications in Anti-Markovnikov additions to terminal alkynes and alkene metathesis. Angewantde Chemie International Edition. 2006;**45**:2176-2203. DOI: 10.1002/anie.200501391, and references therein

[130] Dixneuf HP, Brueau C, Dérien S. Smart ruthenium catalysts for the selective catalytic transformation of alkynes. Pure and Applied Chemistry. 1998;**70**:1065-1070. DOI: 10.1351/pac199870051065

[131] Doucet H, Martin-Veca B, Bruneau C, Dixneuf HP. General synthesis of (Z)-Alk-1-en-1-yl esters via ruthenium-catalyzed *anti*-Markovnikov *trans*-addition of carboxylic acids to terminal alkynes. The Journal of Organic Chemistry. 1995;**60**:7247-7255. DOI: 10.1021/jo00127a033

[132] Doucet H, Höfer J, Bruneau C, Dixneuf HP. Stereoselective synthesis of Z-enol esters catalysed by [bis(diphenylphosphino)alkane]bis(2-methylpropenyl)ruthenium complexes. Journal of the Chemical Society, Chemical Communications. 1993:850-851. DOI: 10.1039/C39930000850

[133] Goossen JL, Paetzolda J, Koleya D. Regiocontrolled Ru-catalyzed addition of carboxylic acids to alkynes: Practical protocols for the synthesis of vinyl esters. Chemical Communication. 2003:706-707. DOI: 10.1039/B211277A

[134] Yi SC, Gao R. Scope and mechanistic investigations on the solvent-controlled regio- and stereoselective formation of enol esters from the ruthenium-catalyzed coupling reaction of terminal alkynes and carboxylic acids. Organometallics 2009;**28**:6585-6592. DOI: 10.1021/om9007357

[135] Melis K, Samulkiewicz P, Rynkowski J, Verpoort F. Ruthenium-catalyzed selective *anti*-Markovnikov *trans* addition of carboxylic acids and tail-to-tail dimerization of terminal alkynes. Tetrahedron Letters. 2002;**43**:2713-2716. DOI: 10.1016/S0040-4039(02)00366-0

[136] Ye S, Leong KW. Regio- and stereoselective addition of carboxylic acids to phenylacetylene catalyzed by cyclopentadienyl ruthenium complexes. Journal of Organometallic Chemistry. 2006;**691**:1117-1120. DOI: 10.1016/j.jorganchem.2005.11.023

[137] Tan TS, Fan YW. Ligand-controlled regio- and stereoselective addition of carboxylic acids onto terminal alkynes catalyzed by carbonylruthenium(0) complexes. European Journal of Organic Chemistry. 2010:4631-4635. DOI: 10.1002/ejic.201000579

[138] Hua R, Tian X. $Re(CO)_5Br$-catalyzed addition of carboxylic acids to terminal alkynes: A high *anti*-Markovnikov and recoverable homogeneous catalyst. The Journal of Organic Chemistry. 2004;**69**:5782-5784. DOI: 10.1021/jo049455n

[139] Jeschke J, Engelhardt BT, Lang H. Ruthenium-catalyzed hydrocarboxylation of internal alkynes. European Journal of Organic Chemistry. 2016:1548-1554. DOI: 10.1002/ejoc.201501583

[140] Kawatsura M, Namioka J, Kajita K, Yamamoto M, Tsuji H, Itoh T. Ruthenium-catalyzed regio- and stereoselective addition of carboxylic acids to aryl and trifluoromethyl group substituted unsymmetrical internal alkynes. Organic Letters. 2011;**13**:3285-3287. DOI: 10.1021/ol201238

[141] Oe Y, Ohta T, Ito Y. Ruthenium catalyzed addition reaction of carboxylic acid across olefins without $\beta$-hydride elimination. Chemical Communications. 2004:1620-1621. DOI: 10.1039/B404229H

[142] Yang C, He C, Gold(I)-catalyzed intermolecular addition of phenols and carboxylic acids to olefins. Journal of the American Chemical Society. 2005;**127**:6966-6967. DOI: 10.1021/ja050392f

[143] Taylor GJ, Whittall N, Hii KK. Copper(II)-catalysed addition of O–H bonds to norbornene. Chemical Communications. 2005:5103-5105. DOI: 10.1039/B509933A

[144] Chen W, Lu J. $In(OTf)_3$-catalyzed intermolecular addition of carboxylic acids and phenols to norbornene under solvent-free conditions. Catalysis Communications. 2007;**8**:1298-1300. DOI: 10.1016/j.catcom.2006.11.031

[145] Choi J, Kohno K, Masuda D, Yasuda H, Sakakura T. Iron-catalysed green synthesis of carboxylic esters by the intermolecular addition of carboxylic acids to alkenes. Chemical Communications. 2008:777-779. DOI: 10.1039/B713951A

[146] Higashi S, Takenaka H, Ito Y, Oe Y, Ohta T. Synthesis of $RuCl_2$(xantphos)L (L = $PPh_3$, $P(OPh)_3$, DMSO) complexes, and their catalytic activity for the addition of carboxylic acids onto olefins. Journal of Organometallic Chemistry. 2015;**791**:46-50. DOI: 10.1016/j.jorganchem.2015.04.016

[147] Rosenfeld C D, Shekhar S, Takemiya A, Utsunomiya M, Hartwig JF. Hydroamination and hydroalkoxylation catalyzed by triflic acid. Parallels to reactions initiated with metal triflates. Organic Letters. 2006;**8**:4179-4182. DOI: 10.1021/ol061174+

[148] Li Z, Zhang J, Brouwer C, Yang C, Reich WN, He C. Brønsted acid catalyzed addition of phenols, carboxylic acids, and tosylamides to simple olefins. Brønsted acid catalyzed addition of phenols, carboxylic acids, and tosylamides to simple olefins. Organic Letters. 2006;**8**:4175-4178. DOI: 10.1021/ol0610035

[149] Perkowski JA, Nicewicz AD. Direct catalytic *anti*-Markovnikov addition of carboxylic acids to alkenes. Journal of the American Chemical Society. 2013;**135**:10334-10337. DOI: 10.1021/ja4057294

Chapter 3

# Applications of Carboxylic Acids in Organic Synthesis, Nanotechnology and Polymers

Aidé Sáenz-Galindo, Lluvia I. López-López,
Fabiola N. de la Cruz-Duran,
Adali O. Castañeda-Facio,
Leticia A. Rámirez-Mendoza,
Karla C. Córdova-Cisneros and
Denisse de Loera-Carrera

Additional information is available at the end of the chapter

http://dx.doi.org/10.5772/intechopen.74654

**Abstract**

Carboxylic acids are versatile organic compounds. In this chapter is presented a current overview of the use of carboxylic acids in a different area as organic synthesis, nanotechnology, and polymers. The application carboxylic acids in these areas are: obtaining of small molecules, macromolecules, synthetic or natural polymers, modification surface of nanoparticles metallic, modification surface of nanostructure such as carbon nanotubes and graphene, nanomaterials, medical field, pharmacy, etc. Carboxylic acids can be natural and synthetic, can be extracted or synthesized, presented chemical structure highly polar, active in organic reactions, as substitution, elimination, oxidation, coupling, etc. In nanotechnology, the use of acid carboxylic as surface modifiers to promote the dispersion and incorporation of metallic nanoparticles or carbon nanostructure, in the area of polymer carboxylic acids present applications such monomers, additives, catalysts, etc. The purpose of this chapter is to emphasize the importance of carboxylic acids in different areas, highlighting the area of organic synthesis, nanotechnology and polymers and its applications.

**Keywords:** carboxylic acids, organic synthesis, nanotechnology, polymers, application

## 1. Introduction

Carboxylic acids are compounds with excellent chemical and physical properties, the most particular characteristics of this type of organic compounds, is their high solubility in polar

solvents, as water, or alcohols, methanol, ethanol, etc. Chemical structure contains a carbonyl function (-C=O) and an hydroxyl group (OH), these groups interact easily with polar compounds, forming bridges of H, obtaining high boiling points. The carbonyl group (C=O) is considered a one of the most functional groups involved in many important reactions. The carboxylic acids are the most important functional group that present C=O.

This type of organic compounds can be obtained by different routes, some carboxylic acids, such as citric acid, lactic acid or fumaric acid are produced from by fermentation, most of these type of carboxylic acids are applied in the food industry. Historically, some carboxylic acids were produced by sugar fermentation. Synthetics route, there are different synthesis reactions such as reactions of oxidation from alcohols in the presence of strong oxidants such as $KMnO_4$, oxidation of aromatic compounds among other routes.

For example, citric acid is a carboxylic acid, can be obtained by different routes, synthetic, enzymatic and naturally occurring, is considered harmless and cheap, used in the food industry, because is non-toxic, has a thermal stability to the 175°C. Bian *et al.*, in 2017, reported the use of citric acid impregnated in porous material for the synthesis of Ni particles. They showed, that the presence of citric acid, is important in the dispersion of the Ni particles when are incorporate in porous materials, thus inhibiting the agglomeration [1].

Derivatives of carboxylic acid, as alkyl halides, esters, and amides, present different and important application in diverse areas. In the case of esters, these are obtained from the reaction between carboxylic acids and alcohols in presence of an acid catalyst usually $H_2SO_4$ with heat, this type of reaction is known as esterification. In the case of the amides, it is obtained in the presence of an amine, may be primary and secondary, with a carboxylic acid, in this reaction also can be used a catalyst and heat to accelerate the reaction.

Due to their chemical and physical characteristics, this type of organic compounds presents innumerable applications in the different areas, such as medicine, pharmacy, organometallic, polymer, nanotechnology, food, among others. Exist different reports, where study carboxylic acid, in the area organic synthesis, in 2008 Lazzarato *et al.*, reported the use of a carboxylic acid, salicylic acid type "aspirin-like", molecule obtained through a novel approach, where the phenol reaction to nitrooxy-acyl, this molecule present pharmaceutical properties [2].

In nanotechnology, the carboxyl acid, present in different applications, in 2016, Sáenz *et al.*, reported the use of organic carboxylic acids: tartaric acid, maleic acid, and malic acid, assisted the surface modification of multiple wall carbon nanotubes (MWCNTs) by ultrasonic radiation, with applications in the production of polymer nanomaterials. Finding that the modification with this type of carboxylic acids favors the dispersion of MWCNTs in the polymer matrix [3].

In the polymer area, also the carboxylic acids present important applications, in 2017 Oguz *et al.*, reported the obtain of "green composites" of Poli(lactic acid) and 10 wt % waste cellulose fibers, demonstrating that is easy, economical and sustainable its obtaining, this "green composites" presented improved tensile and impact properties [4]. Yasa *et al.*, reported the synthesis and characterization of polyol esters from iso-undecenoic and iso-undecanoic acids, using montmorillonite K10 clay as a catalyst, in presence of deionized water at a temperature of 250°C in an autoclave. This type of polymer presented applications like lubricant properties and good oxidation stability [5].

In general, the carboxylic acids presents applications in different areas, the propose of this chapter is to show a general panorama, about the applications of carboxylic in organic synthesis, nanotechnology, and polymer.

## 2. Use of carboxylic acids

### 2.1. Organic synthesis

The use of carboxylic acids in organic synthesis is a very wide area and the chemical transformations of this group to another have made it a very versatile functional group. These chemical transformations have seen improvement when they carry out through Green chemistry processes.

One of the methods to aim for energy efficiency (one of the principles of Green Chemistry) is to make reactions under microwave irradiation. The first report by this methodology was an esterification reaction with carboxylic acids and alcohols obtaining high yields in a short reaction time [6].

In addition to the esterification, the amidation reaction by a transformation of the carboxylic acids is also important because of the new covalent bond formed. This bond is of great importance because it can be found in a wide variety of molecules both in natural products and in small molecules with pharmacological activity [7–9]. Therefore, the development of new direct amidation reactions is important [10, 11]. One of those direct methodologies of amide formation is by reacting amines with carboxylic acids using toluene as a solvent [12] or using radiofrequency heating under neat conditions [13]. However this reagent-free pathway has limitations in the substrate scope.

On the other hand, Lanigan *et al.* reported a methodology in which they used simple borate esters that are efficient reagents for the direct synthesis of amides, using a variety of carboxylic acids and amines [14]. This reaction can be carried out openly to the air under acetonitrile reflux. The amidation product can be purified in a very simple way, in most cases, it only needs a simple filtration procedure using commercially available resins giving excellent yields (**Figure 1**). <http://pubs.acs.org/doi/abs/10.1021%2Fjo400509n>.

Another methodology in which microwave irradiation was used in the amidation reaction has proven to be an efficient synthesis of amides [15–18]. For example, Ojeda-Porras *et al.* [19] described a green methodology for the direct amidation (**3**) of carboxylic acids (**1**) and amines (**2**) using silica gel as a solid support (**Figure 2**).

In addition to the typical use of carboxylic acids in transformations other carbonyl groups, they can be used as a substrate in multicomponent reactions such as the well-known Ugi [20, 21]

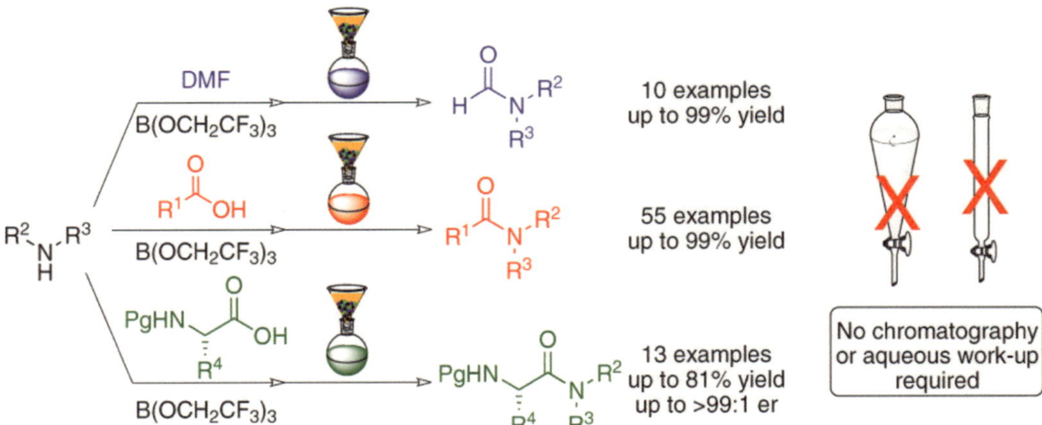


**Figure 1.** Carboxylic acids were used to obtain a variety of amides [12].

and Passerini [22, 23] reactions. An example is the reaction between an oxazolidine **4**, an isocyanide **5** and a carboxylic acid **6** to provides *N*-acyloxy ethylamino acid derivatives (**7**) (**Figure 3**) [24] so that it can be complemented with those structures that are produced by the well-established Ugi and Passerini reactions and they allow the generation of a chemical libraries.

As well, carboxylic acids can be used in organocatalytic reactions. As is in the case, carboxylic acid **11** can be used directly to obtain the $\alpha$-hydroxy phosphonates **12**. The reaction was carried out in a simple way with a variety of aldehydes **8** and ketones **9** with trimethylphosphite **10** in the presence of catalytic amounts of pyridine 2,6-dicarboxylic acid **11** in water as a solvent. This generates a low cost and environmentally friendly methodology (**Figure 4**) [25].

Another example of using carboxylic acids as catalysts is by incorporating them into the structure of ionic liquids (IL). Jahani *et al.* described the condensation of 1,8-dioxo-octahydroxanthene **16** with 5,5-dimethyl-1,3-cyclohexanedione (**13**) and aldehyde derivatives **14** using carboxylic acid functionalized IL (**15**) under microwave irradiation (**Figure 5**) [25]. <https://creativecommons.org/licenses/by-nc-nd/4.0/> provides and offers the advantages of using IL and microwave irradiation are making the reaction an efficient and eco-friendly procedure.

Carboxylic acids present important applications in the pharmaceutical area, due to their chemical structure. Different methods have been developed for their detection, in medicines, in cosmetics, in food additives, etc. In 2015, Soham *et al.*, reported a selective chromogenic system, which not only can discriminate maleic acid vs. fumaric acid but can also differentiate

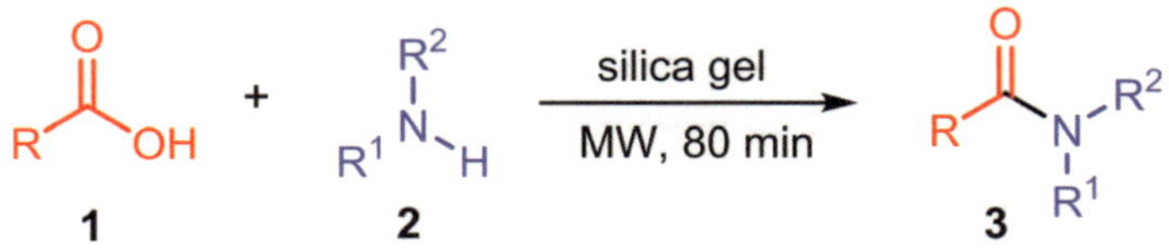

**Figure 2.** Amidation reaction via carboxylic acids and amines [17].

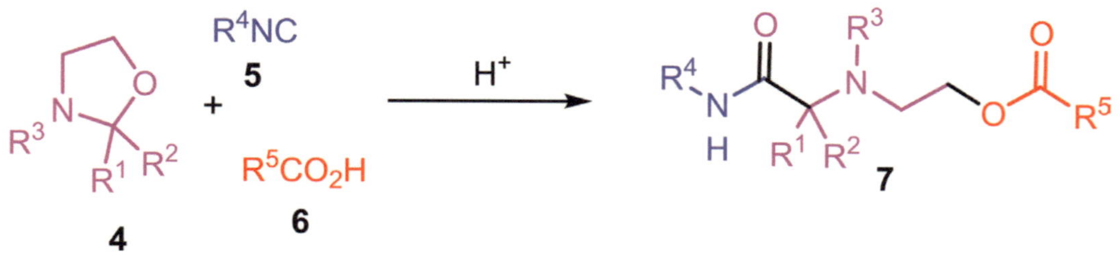

Figure 3. Use of carboxylic acid in a multicomponent reaction [24].

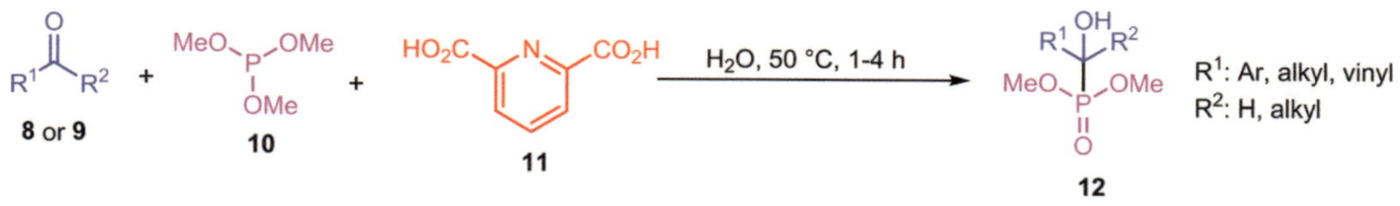

Figure 4. Carboxylic acid as a catalyst [25].

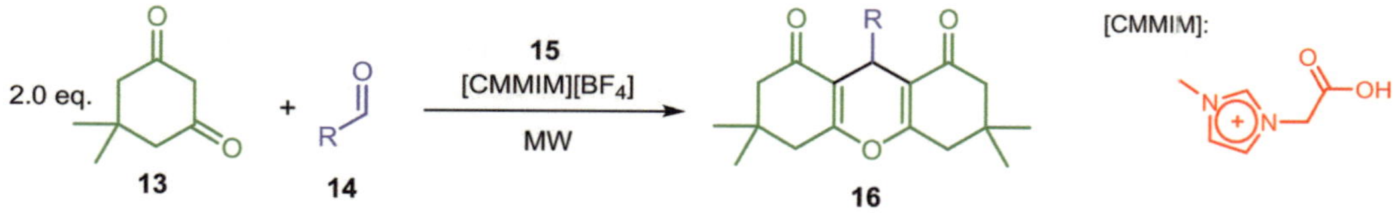

Figure 5. Carboxylic acid in ionic liquid in the multicomponent reaction [26].

maleic acid among diverse carboxylic acids. This method uses sharp colorimetric, as well as fluorogenic responses in both physiological conditions and food additives. The detection of this type of organic acids is very important. For example, maleic acid plays an important role in the organism because is a Krebs cycle inhibitor whereas, fumarate is produced in the Krebs cycle. Excessive consumption of maleic acid can cause different kidney diseases [27].

### 2.2. Nanotechnology

Recently carboxylic acids have been studied extensively, due to their important applications in the petrochemical, food industry, dyes, stabilizers and currently in nanotechnology [28]. This type of acid has become very important because it has been considered as a green solvent as part of the Eutectic Deep Solvents (DES), studied in 2003 by Abbott *et al.* [29] which are obtained by a mixture of hydrogen bond acceptor (such as a quaternary ammonium salt) and a hydrogen bond donor species (proton donating species). One of the most common acids is carboxylic acids, belonging to green solvents because they are not highly toxic and inexpensive [30].

One of the most important applications today of the carboxylic acids is the surface modification of the nanoparticles, this because during the synthesis of the nanoparticles by any methodology these tend to agglomerate due to the van der Waals forces and the absence of repulsive forces. In addition, oxidation at the surface of the nanoparticles causes instability which leads to aggregation [31]. One of the strategies to avoid this problem is to protect the colloidal particles with a passivating or stabilizing agent, which associates with the surface of the nanoparticles to keep them suspended, and therefore to prevent their aggregation [32]. In addition to acting as stabilizers, carboxylic acids also influence the solubility, reactivity, size and shape of nanoparticles [33]. Among other surface modifiers, the most used agents to stabilize the surface of the nanoparticles are polyvinylpyrrolidone (PVP), chitosan, starch, cellulose, and gelatin. Carboxylic acids serve as stabilizers for the preparation of nanoparticles, due to the carboxyl group provides coordination to the nanoparticles surface and therefore they stabilize. An example of this is oleic acid, which is widely used for the stabilization of nanoparticles, as well as for controlling the size and morphology of nanoparticles [34]. In this sense in 2008 Yang *et al.*, obtained nanoparticles of $Fe_3O_4$ with sizes of 6–30 nm using oleic acid as surfactant or surfactant, this group of researchers functionalized the nanoparticles with carboxylic acids obtaining acid catalysts for the hydrolysis of carbohydrates, being able to observe that the acid functionalization can have large advantages in producing more active catalysts and thus an application in green processes [35]. On the other hand, Cabello *et al.* were able to surface functionalize multiwall carbon nanotubes using acetic acid and aniline, assisted by ultrasound, demonstrating the hydrophilic behavior of carbon nanotubes [36].

Hojjati *et al.*, modified $TiO_2$ nanoparticles with carboxylic acid followed by polymerization with acrylic acid to obtain a well-dispersed nanoparticle in polyacrylic acid [37]. Armenalo used the carboxylic acid as a solvent to obtain CuS particles, finding several advantages with the use of this solvent because it favors the hydrolysis of the C-S bond thus producing a fast CuS supersaturation and a high speed of nucleation, favoring the growth of the particles and prevents agglomeration of the particles [38]. Qu *et al.* studied the chemical modification of nanoparticles of $TiO_2$ with carboxylic acids by the solvothermal method finding improvement in the photovoltaic performance of the $TiO_2$ nanoparticles despite being coated with carboxylic acid [39].

In other studies, it has been shown that the use of carboxylic acid as a surface modifier of nanoparticles can be easily redispersed in diverse matrix or solvents and improves properties as antibacterial activity [40]. In recent years, inorganic nanoparticles have been widely studied due to the excellent properties that they provide, due to their large surface area, emphasizing applications such as optical, catalytic, electrical, sensing, transport, magnetic, thermal conductivity, electromagnetic. These properties are the result of the large surface area that they possess. Metals such as gold, silver, palladium, and copper have been used to make inorganic nanoparticles of various shapes and sizes [41, 42]. The procedure and the conditions of synthesis of the nanoparticles directly influence its shape and size.

A wide variety of methods have been developed to synthesize metallic nanoparticles with different morphologies, as, nanotubes, nanodisks, nanofibers and others. In general, these procedures are classified into three groups: chemical methods, physical and biological methods.

Chemical methods are the most used, due to their ease of climbing [43]. The chemical reduction of metal salts in solution is the most commonly used method [44]. According to Slistan, the chemical reduction allows adequate control of the size, size distribution, and shape of nanoparticles [45].

In aqueous systems, the reducing agent is added or generated in situ, among the most commonly used reducing agents are sodium borohydride, hydrazine, and dimethylformamide, however in recent year's nontoxic and equally effective substances have been used, such as sodium citrate and glucose [46]. On the other hand, in non-aqueous systems, the solvent can also act as a reducing agent [47], such solvents can be alcohols such as polyethylene glycol, glycerol, and ethylene glycol, through which colloidal nanoparticles are obtained [48]. The advantage of these systems is that addition of reducing agents is not required, even synthesizing at room temperature [49]. Physical methods include electrochemical methods, laser ablation, thermolysis, microwave irradiation, and sonochemistry. For example, the thermolysis method involves the decomposition of solids at high temperature; through this process it is possible to obtain particles smaller than 5 nm [43].

Biologicals methods are developed using a metal salt and a reducing agent, which may be microorganisms, enzymes or plant extracts [49].

However, in general, the nanoparticles obtained by any methods tend to agglomerate due to the absence of repulsive forces and the forces of attraction of van der Waals [47]. Another common characteristic of metallic nanoparticles (Ag, Cu, etc.) is the oxidation of the surface causing instability of the nanoparticles and deriving in the aggregation of the same [31]. As a strategy to avoid these two major problems, the nanoparticles are modified superficially, to keep them stable in an aqueous solution and therefore to prevent their aggregation [50]. In addition to acting as stabilizers, surface modifiers also influence the solubility, reactivity, size, and shape of nanoparticles [33]. The surface modification of inorganic nanoparticles has today attracted attention, because they can be incorporated perfectly into some polymer, ceramic or metal matrix, improving the interaction between the two phases (the dispersed phase refers to the inorganic nanoparticles and the continuous phase is referred to the metallic, ceramic or polymer phase) formed nanocomposite or hybrid materials, combining the properties of the two phases [51, 52].

There are several types of surface modification such as chemical treatments, synthetic polymer grafts, the ligand exchange technique, among others; modification by chemical treatments may include reactions with metal alkoxides, carboxylic acid epoxides, silane coupling agents, among others; this being a convenient method to avoid the agglomeration of the nanoparticles and thus improve the dispersion [51], to achieve the surface modification of nanoparticles has been used ultrasonic irradiation as a technique that obeys the principles of green chemistry, because it reduces the use of solvents as well as energy. Another benefit of this technique is that the reaction mixture can be heated rapidly and uniformly, resulting in a shorter reaction time and a complete interaction avoiding the production of by-products [53].

Nanoparticles can be surface modified with adsorption of polymers on the surface, is one of the simplest methods to improve the dispersion of nanoparticles in aqueous systems. Nanoparticles with a hydrophilic behavior can be dispersed in polar solvents by modifying them with anionic or cationic polymeric dispersants. These dispersants generate repulsive forces between the particles and increase the dispersibility of the nanoparticles. In this case, the polycarboxylic acids, trioctylphosphine (TOPO), oleic acid and amines are examples of surfactants that modify nanoparticles of $Al_2O_3$, $Fe_2O_3$, $TiO_2$ and that can even be added during the synthesis of the nanoparticles [54].

The nanoparticles can also be functionalized with organic molecules with biological functions, e.g., lipids, vitamins, peptides, and sugars, in addition to other macromolecules such as proteins, enzymes, DNA and RNA. The combination of inorganic nanoparticles and biomolecules allows the use of these in biological systems because they combine unique properties for applications such as molecular recognition [55].

Another very important methodology found within the principles of green chemistry is the use of plasma to modify the surface of nanoparticles. Plasma surface modification is an effective and economical surface treatment technique for many nanoparticles.

This is a relatively simple, fast and dry method which has been used to modify the surface chemistry of different nanoparticles. It has also proved to be a versatile method since it allows the use of gases and/or organic molecules, whereby the surface modification can be adjusted to specific requirements, in other words, a particular functional group can be chosen.

This technique was originally implemented to modify the polymer surfaces, but in recent years has been used for the treatment of different nanoparticles [56]. The principle of the plasma polymerization technique is the creation of ionized and radical molecules by the bombardment of an electromagnetic field. These molecules and radicals can react with the surface of the substrate by erosion, removal, and deposition. As a result, the surface properties of the substrate will be modified.

Several organic molecules such as acrylic acid, pyrrole and styrene have been used in plasma to modify nanoparticles of zinc oxide, alumina and carbon nanofibers, respectively [57]. By this process it is possible to form a thin layer ranging from 1 to 3 nm in thickness.

The analytical methods used for characterizing the obtained nanoparticles are transmission electron microscopy (TEM) for size and shape determination, dynamic light scattering, fluorescence spectroscopy. Furthermore, binding of ligand molecules to metal nanoparticles can be probed by surface-enhanced Raman scattering.

Silver nanoparticles are recognized with antifungal and antibacterial properties and have applications in biosensing, antivirals against HIV-I, in water purification systems and paint products. In case of Au nanoparticles, these are used in cancer diagnosis and therapy, antiviral and antibacterial, MRI, biosensing applications. Magnetic nanoparticles have antimicrobial properties and are used in biomedicine such as drug delivery, magnetic resonance, and cancer treatment. The uses of $TiO_2$ nanoparticles that are also antimicrobial are in skin care products, nanomedicine, photocatalysis, gas sensor, wastewater to eliminate organic and inorganic pollutants. ZnO nanoparticles also have antimicrobial and anticancer properties, used in cosmetics, medical fillers, electronic and optoelectronic devices and gas sensors. The main applications of the nanoparticles of $Al_2O_3$ are in drug delivery, in membranes to eliminate pathogenic microorganisms, antimicrobial applications, removal of heavy metals from wastewater, in catalysis and in gas separation processes. On the other hand, the applications of $SiO_2$ nanoparticles are in biosensing, also serve as a carrier for antimicrobial applications, drug delivery, and tissue engineering [53].

Carboxylic acids are currently used as surface modifiers in carbon-based nanostructure: carbon nanotubes, single wall and multiple walls, graphene, nanofibers, etc., with the purpose of improving the dispersion in polar solvents such water, ethanol, methanol, ethyl acetate, etc. This type of superficial modifications can be carried out through different alternatives making use of green chemistry, which recommend the use of sustainable activation energies such as ultrasound, microwaves, plasma, that helps to reduce energy consumption and decrease the time of reaction in the surface modification.

### 2.3. Polymer

The carboxylic acids present applications in the obtaining of polymers, acting as monomers, additives, initiators, catalysts, dopants, etc. Currently, an area of great interest is the production of acidic polymers, with different applications, for example in electronic area required that present characteristics such as electron donors, high solubility in aqueous solvents, etc. [58].

As additive the carboxylic acids have been studied. In 2017, it was reported the study of a series of linear carboxylic acids whit different chain lengths of 6 trans-2-hexanoic acid carbon atoms (CA-6), trans-2-decanoic acid (CA-10), 9-tetradecanoic acid (CA-14), used as halogen-free additives-solvent, considered a sustainable and viable process useful for the production of polymeric films whit optical properties, whit potential application in solar cells. The conclusion of the study showed that increasing the length of the carboxylic acid chain changes the topology of the polymeric film [59].

Recent studies reported the use of acids polymers as dopants, to access and stabilize the electrically conducting states of conducting. The acids polymers are used for replace small-molecules acids as: poly(acrylic acid), poly(styrenesulfonate) (PSS), and poly(2-acrylamido-2-methyl-1-propanesulfonic acid), (2-acrylamido-2-methyl-1-propanesulfonic acid) (PAAMPSA), because its chemical nature, helps to stabilize the conduction of the polymers, in conclusion, the presence of an excess of carboxylic acid in the chemical structure promotes good dispersion, thus stabilizing the electric nature of the doped polymers [60].

In 2012 Shi *et al.*, reported the obtaining of two new blue transmissive donor-acceptor electrochromic polymers: a polymeric material synthesized by alternating copolymerization route and another random copolymerization, demonstrating that this type of polymeric materials having characteristic as higher water solubility. The results of this study, show that these polymeric acids, present electrochromic properties [61].

Recently Mohammadifar *et al.*, reported the reaction of cationic polymerization at room temperature and solvent free to obtain polyglycerol, the polymerization is classified as green polymerization, in this type of polymerization, citric acid participates as a donor of proton, promoting polymerization, it was reported that this type of polymerization does not require a purification process which indicates that the process is very sustainable, these types of polymers may be potential candidates for biomedical applications [62].

There are different types of polymer acids, natural or synthetic, for example, polylactic acid presents applications as polymeric acid antibacterial, due to its physical and chemical characteristics, the polylactic acid can be modified or incorporated whit other polymer matrices and form composites, it also presents high biodegradability and biocompatibility, which makes it a sustainable material [63–65].

Poly (methacrylic acid), is a peculiar material, its chemical structure consists of a polar backbone where the carboxylic acid is and rest of the structure is no polar. In 2017, Robin *et al.*, reported the study of behavior rheological of concentration of the poly (methacrylic acid) in aqueous solutions and influence in viscosity [66], the purpose of this study is know the physic-chemical of the polymeric material and to be able to formulate different composites, concluding that the poly (methacrylic acid) in aqueous solutions present a rheological behavior controlled by the balance between in the different interactions intramolecular as hydrogen-bonding [66]. Also the poly (methacrylic acid) is considered a biodegradable and sustainable polymeric material, used for the obtaining of biomaterials. Poly (methacrylic acid) brushes are highly susceptible to swelling in aqueous solution due to ionization of carboxylic groups presents in their chemical structure [67].

In general, the applications of carboxylic acid in the polymer are due to the presence of a polar group that helps solvate in aqueous systems, facilitating its processing.

## 3. Conclusion

In conclusion, the carboxylic acids have been widely applied in different areas, highlighting organic synthesis, nanotechnology and polymers, have been used in basic applications until relevant application. In organic synthesis, carboxylic acids can act as an organic substrate, reagents in reactions "*one step*" which are considered "green reactions" in some cases, also are used as catalysts, presents activity in substitution, addiction, condensation, polymerizations and copolymerization reactions. Actually, the research in organic synthesis is directed toward green reactions, easy, fast, economical, sustainable processes, the carboxylic acids are reagents that present a high reactivity, due to the chemical nature of the carbonyl group. In nanotechnology, the carboxylic acids are used as modifiers substrate in surface modifications of the nanostructure of carbon: CNT's or graphene, the proposal to obtain a polar surface thus improving the dispersion of the nanostructure of carbon in different polar matrices. The chemical nature of the carboxylic acids allows to carry out this type of application. Currently, in the area of polymeric materials, the carboxylic acids have been demonstrated that present important application as soluble polymer, degradation polymeric processes, obtain of composites and hybrid, a hydrophilic polymer, etc. It is considered that the application of the carboxylic acids is due to natural chemical, specifically by hydrophilic characteristic provided by the functional group. The research on the use of carboxylic acids, is very interesting, in this chapter were described some recent report. However, there are still vast challenges within this field that remain to explored.

## Acknowledgements

The authors acknowledge the financial support from project Modification of Carbon Nanotubes with Nitrogen Organic compounds obtained by Green Methodologies in the Thematic Collaboration Network: Green Organic Synthesis with Application in Nanotechnology, funded by PRODEP-México.

## Author details

Aidé Sáenz-Galindo[1]*, Lluvia I. López-López[1], Fabiola N. de la Cruz-Duran[1], Adali O. Castañeda-Facio[2], Leticia A. Rámirez-Mendoza[1], Karla C. Córdova-Cisneros[2] and Denisse de Loera-Carrera[3]

*Address all correspondence to: aidesaenz@uadec.edu.mx

1 Department of Organic Chemistry, Autonomous University of Coahuila, Saltillo, México

2 Department of Science and Technology of Polymer, Autonomous University of Coahuila, Saltillo, México

3 Autonomous University of San Luis Potosí, San Luis Potosí, México

## References

[1] Bian Z, Xin Z, Meng X, Tao M, Lv Y, Gu J. Effect of citric acid on the synthesis of CO methanation catalysts with high activity and excellent stability. Industrial and Engineering Chemistry Research. 2017;**56**:2383-2392. DOI: 10.1021/ac.iecr.6b04027

[2] Lazzarato L, Donnola M, Rolando B, Marini E, Cena C, Coruzzi G, Guaita E, Morini G, Fruttero G, Gasco A, Biondi S, Ongini E. Searching for new NO-donor aspirin-like molecules: A new class of nitrooxy-acyl derivatives of salicylic acid. Journal of Medicinal Chemistry. 2008;**51**:1894-1903. DOI: 10.1021/jm701104f

[3] Sáenz A, Rodríguez K, Rubio M, Bermúdez L, Ramírez L, Avila C, Jiménez R. Assisted surface modification with ultrasonic energy of carbon nanotubes with malic acid, malonic acid and tartaric acid. Chemistry Advance. 2016;**11**:47-52

[4] Oguz O, Bilge K, Simsek E, Citak MK, Wis AA, Ozkoc G, Menceloglu YZ. High-performance green composites of poly(lactic acid) and waste cellulose fibers prepared by high-shear thermokinetic mixing. Industrial and Engineering Chemistry Research. 2017;**56**:8568-8579

[5] Yasa SR, Krishnasamy S, Singh KR, Penumarthy V. Synthesis and characterization of iso-undecenoic and isoundecanoic acids based polyol esters. Industrial and Engineering Chemistry Research. 2017;**56**:7423-7433

[6] Gedye R, Smith F, Westaway K, Ali H, Baldisera L, Laberge L, Rousell J. The use of microwave ovens for rapid organic synthesis. Tetrahedron Letters. 1986;**27**:279-282. DOI: 10.1016/S0040-4039(00)83996-9

[7] Roughley SD, Jordan AM. The medicinal chemist's toolbox: An analysis of reactions used pursuit of drug candidates. Journal of Medicinal Chemistry. 2011;**54**:3451-3479. DOI: 10.1021/jm200187y

[8] Carey JS, Laffan D, Thomson C, Williams MT. Analysis of the reactions used for the preparation of drug candidate molecules. Organic & Biomolecular Chemistry. 2006;**4**:2337-2347. DOI: 10.1039/B602413K

[9] Constable DJC, Dunn PJ, Hayler JD, Humphrey GR, Leazer JJL, Linderman RJ, Lorenz K, Manley J, Pearlman BA, Wells A, Zaks A, Zhang TY. Key green chemistry research areas-a perspective from pharmaceutical manufacturers. Green Chemistry. 2007;**9**:411-420. DOI: 10.1039/B703488C

[10] Ghose AK, Viswanadhan VN, Wendoloski JJJ. A knowledge-based approach in designing combinatorial or medicinal chemistry libraries for drug discovery. 1. A qualitative and quantitative characterization of known drug databases. Journal of Combinatorial Chemistry. 1999;**1**:55-68. DOI: 10.1021/cc9800071

[11] Amarnath L, Andrews I, Bandichhor R, Bhattacharya A, Dunn P, Hayler J, Hinkley W, Holub N, Hughes D, Humphreys L, Kaptein B, Krishnen H, Lorenz K, Mathew S, Nagaraju G, Rammeloo T, Richardson P, Wang L, Wells A, White T. Green chemistry articles of interest to the pharmaceutical industry. Organic Process Research and Development. 2012;**16**:535-544. DOI: 10.1021/op300068d

[12] Allen CL, Chhatwal AR, Williams JM. Direct amide formation from unactivated carboxylic acids and amines. Journal of Chemical Communications. 2012;**48**:666-668. DOI: 10.1039/C1CC15210F

[13] Houlding TK, Tchabanenko K, Rahman Md. T, Rebrov EV. Direct amide formation using radiofrequency heating. Organic and Biomolecular Chemistry. 2013;**11**:4171-4177. DOI: 10.1039/C2OB26930A

[14] Lanigan RM, Starkov P, Sheppard TD. Direct synthesis of amides from carboxylic acids and amines using $B(OCH_2CF_3)_3$. The Journal of Organic Chemistry. 2013;**78**:4512-4523. DOI: 10.1021/jo400509n. http://pubs.acs.org/doi/abs/10.1021%2Fjo400509n (Notice: for further permissions related to the material should be directed to the ACS)

[15] Lukasik N, Wagner-Wysiecka E. A review of amide bond formation in microwave organic synthesis. Current Organic Synthesis. 2014;**11**:592-604. DOI: 10.2174/15701794 11666140321180857

[16] Varma RS, Naicker KP. Solvent-free synthesis of amides from non-enolizable esters and amines using microwave irradiation. Tetrahedron Letters. 1999;**40**:6177-6180. DOI: 10.1016/S0040-4039(99)01209-5

[17] Yaragorla S, Singh G, Lal Saini P, Reddy MK. Microwave assisted, Ca (II)-catalyzed Ritter reaction for the green synthesis of amides. Tetrahedron Letters. 2014;**55**:4657-4660. DOI: 10.1016/j.tetlet.2014.06.068

[18] Perreux L, Loupy A, Volatron F. Solvent-free preparation of amides from acids and primary amines under microwave irradiation. Tetrahedron. 2002;**58**:2155-2162. DOI: 10.1016/S0040-4020(02)00085-6

[19] Ojeda-Porras A, Hernández-Santana A, Gamba-Sánchez D. Direct amidation of carboxylic acids with amines under microwave irradiation using silica gel as a solid support. Green Chemistry. 2015;**17**:3157-3163. DOI: 10.1039/C5GC00189G

[20] Ugi I, Meyr R, Fetzer U, Steinbruckner C. UGI. München: Versuche mit Isonitrilen. Angewandte Chemie. 1959;**71**:386. DOI: 10.1002/ange.19590711110

[21] Ugi I. The $\alpha$-addition of immonium ions and anions to isonitriles accompanied by secondary reactions. Angewandte Chemie International Edition. England. 1962;**1**:8-21. DOI: 10.1002/anie.196200081

[22] Passerini M. Sopra gli Isonitrili (I). Composto del p-Isonitrilazobenzolo con Acetone ed Acido Acetico. Gazzetta Chimica Italiana. 1921;**51**:126-129

[23] Passerini M. Sopra gli Isonitrili (II). Composti con Aldeidi o con Chetoni ed Acidi Organici Monobasici. Gazzetta Chimica Italiana. 1921;**51**:181-188

[24] Diorazio LJ, Motherwell WB, Sheppard TD, Waller RW. Observations on the reaction of N-alkyloxazolidines, isocyanides and carboxylic acids: A novel three-component reaction leading to N-acyloxyethylamino acid amides. Observations on the reaction of N-alkyloxazolidines, isocyanides and carboxylic acids: A novel three-component reaction leading to N-acyloxyethylamino acid amides. Synlett. 2006;**14**:2281-2283. DOI: 10.1055/s-2006-950413

[25] Jahani F, Zamenian B, Khaksar S, Taibakhsh M. Pyridine 2,6-dicarboxylic acid as a bifunctional organocatalyst for hydrophosphonylation of aldehydes and ketones in water. Synthesis. 2010;**19**:3315-3318. DOI: 10.1055/s-0030-1257866

[26] Dadhania AN, Patel VK, Raval DK. Ionic liquid promoted facile and green synthesis of 1,8-dioxo-octahydroxanthene derivatives under microwave irradiation. Journal of Saudi Chemical Society. 2014:S163-S168. DOI: 10.1016/j.jscs.2013.12.003. https://creativecommons.org/licenses/by-nc-nd/4.0/>

[27] Soham S, Chirantan K, Gopal D. Colorimetric and fluorometric discrimination of geometrical isomers maleic acid vs fumaric acid) with real-time detection of maleic acid in solution and food additives. Analytical Chemistry. 2015;**87**:9002-9008. DOI: 10.1021/acs.analchem.5b02202

[28] Pai Z, Selivanova N, Oleneva P, Berdnikova P, Beskopyl'nyi A. Catalytic oxidation of alkenes with hydrogen peroxide to carboxylic acids in the presence of peroxopolyoxotungstate complexes. Catalysis Communications. 2017;**88**:45-49. DOI: 10.1016/j.catcom.2016.09.019

[29] Abbott A, Capper G, Davies D, Rasheed R, Tambyrajah V. Novel solvent properties of choline chloride/urea mixtures. Chemical Communications. 2003;**0**:70-71. DOI: 10.1039/B210714G

[30] Teles A, Capela E, Carmo R, Coutinho J, Sivestre A, Freire M. Solvatochromic parameters of deep eutectic solvents formed by ammonium-based salts and carboxylic acids. Fluid Phase Equilibria. 2017;**448**:15-21. DOI: 10.1016/j.fluid.2017.04.020

[31] Panigrahi S, Praharaj S, Basu S, Ghosh SK, Jana S, Pande S, Vo-Dinh T, Jiang H, Pal T. Self-assembly of silver nanoparticles: Synthesis, stabilization, optical properties, and application in surface-enhanced Raman scattering. The Journal of Physical Chemistry. B. 2006;**110**:13436-13444. DOI: 10.1021/jp062119l

[32] Xuping S, Yonglan L. Preparation and size control of silver nanoparticles by a thermal method. Materials Letters. 2005;**59**:3847-3850. DOI: 10.1016/j.matlet.2005.07.021

[33] Sarkar A, Kapoor S, Mukherjee T. Synthesis of silver nanoprisms in formamide. Journal of Colloid and Interface Science. 2005;**287**:496-500. DOI: 10.1016/j.jcis.2005.02.017

[34] Hosseini M, Parchegani F, Alavi S. Carboxyilic acid effects on the size and catalytic activity of magnetite nanoparticles. Journal of Colloid and Interface Science. 2015;**437**:1-9. DOI: 10.1016/j.jcis.2014.08.056

[35] Yang H, Ogawa T, Hasegawa D, Takahashi M. Synthesis and magnetic properties of monodisperse magnetite nanocubes. Journal of Applied Physics. 2008;**103**:7-10. DOI: 10.1063/1.2833820

[36] Cabello C, Sáenz A, Pérez C, López L, Barajas L, Cantú L, Ávila C. Modification of multiwall carbon nanotubes (MWNTC's) using acetic acid and aniline by ultrasonic radiation. Revista Latinoamericana de Metalurgia y Materiales. 2015;**35**:27-33

[37] Hojjati B, Sui R, Charpentier. Synthesis of $TiO_2$/PAA nanocomposite by RAFT polymerization. Polymer. 2007;**48**:5850-5858

[38] Armelao L, Camozzo D, Gross S, Tondello E. Synthesis of copper sulphide nanoparticles in carboxylic acids as solvent. Journal of Nanoscience and Nanotechnology. 2006;**6**:401-408

[39] Qiyun Q, Hongwei G, Ruixiang P, Qi C, Xiaohong G, Fanqing L, Mingtai W. Chemically binding carboxylic acids onto $TiO_2$ nanoparticles with adjustable coverage by solvothermal strategy. Langmuir. 2010;**26**:9539-9546. DOI: 10.1021/la100121n

[40] Cash BM, Wang L, Benicewicz BC. The preparation and characterization of carboxylic acid-coated silica nanoparticles. Journal of Polymer Science part a: Polymer Chemistry. 2012;**50**:2533-2540. DOI: 10.1002/pola.26029

[41] Sárkány A, Papp Z, Sajó I, Sachay Z. Unsupported Pd nanoparticles prepared by $\gamma$-radiolysis of $PdCl_2$. Solid State Ionics. 2005;**176**:209-215. DOI: 10.1016/j.ssi.2004.06.008

[42] Narayanan K, Sakthivel N. Coriander leaf mediated biosynthesis of gold nanoparticles. Material Letters. 2008;**62**:4588-4590. DOI: 10.1016/j.matlet.2008.08.044

[43] Poole C, Owens F. Introduction to Nanotechnology. Hoboken ed. John Wiley & Sons Inc; 2003. p. 402. DOI: 10.1557/mrs2005.114

[44] Masala O, Seshadri R. Synthesis routes for large volumes of nanoparticles. Annual Review of Materials Research. 2004;**34**:41-48. DOI: 10.1146/annurev.matsci.34.052803.090949

[45] Slistan A, Herrera R, Rivas J, Avalos M, Castillon F, Posadas A. Synthesis of silver nanoparticles in a polyvinylpyrrolidone (PVP) paste, and their optical propiertes in a film and in ethylene glicol. Material Research Bulletin. 2008;**43**:90-96. DOI: 10.1016/j.materresbull.2007.02.013

[46] Raveendran P, Fu J, Wallen S. A simple and "green" method for the synthesis of Au, Ag, and Au-Ag alloy nanoparticles. Green Chemistry. 2006;**8**:34-38. DOI: 10.1039/B512540E

[47] Bradley JS, Schmid G.. Conclusions and perspectives. In: Gunter Schumid, editor. Nanoparticles: From Theory to Applications. EUA: Wiley-VCH Verlag GmbH & Co. KGaA; 2004. DOI: 10.1002/3527602399.ch7

[48] Luo C, Zhang Y, Zeng X, Zeng Y, Wang Y. The role of poly(ethylene glycol) in the formation of silver nanoparticles. Journal of Colloid and Interface Science. 2005;**288**:444-448. DOI: 10.1016/j.jcis.2005.03.005

[49] Mohamed A, Mohamed S, Aziza E, Mosaad A. Antioxidant and antibacterial activity of silver nanoparticles biosynthesiized using Chenopodium murale leaf extract. Journal of Saudi Chemical Society. 2014;**18**:356-363. DOI: 10.1016/j.jscs.2013.09.011

[50] Sun X, Luo Y. Preparation and size control of silver nanoparticles by a thermal method. Materials Letters. 2005;**59**:3847-3850. DOI: 10.1016/j.matlet.2005.07.021

[51] Kango S, Kalia S, Celli A, Njuguna J, Habibi Y, Kumar R. Surface modification of inorganic nanoparticles for development of organic–inorganic nanocomposites — A review. Progress in Polymer Science. 2013;(**8**):1232-1261. DOI: 10.1016/j.progpolymsci.2013.02.003

[52] Hethnawi A, Nassar N, Vitale G. Preparation and characterization of polyethylenimine-functionalized pyroxene nanoparticles and its application in wastewater treatment. Colloids and Surfaces, A: Physicochemical and Engineering Aspects. 2017;**525**:20-30. DOI: 10.1016/j.colsurfa.2017.04.067

[53] He J, Zhou L, Yang L, Zou L, Xiang J, Dong S, Xiaochao Y. Modulation of the surface structure and catalytic properties of cerium oxide nanoparticles by thermal and microwave synthesis techniques. Applied Surface Science. 2017;**402**:469-477. DOI: 10.1016/j.apsusc.2017.01.149

[54] Sato K, Kondo S, Tsukada M, Ishigaki T, Kamiya H. Influence of solid fraction on the optimum molecular weight of polymer dispersants in aqueous $TiO_2$ nanoparticle suspensions. Journal of the American Ceramic Society. 2007;**90**:3401-3406

[55] Sperling RA, Parak WJ. Surface modification, functionalization and bioconjugation of colloidal inorganic nanoparticles. Philosophical Transactions of the Royal Society A. 2010;**368**:1333-1383. DOI: 10.1098/rsta.2009.0273

[56] Shi D, Wang SX, Van Ooij WJ, Wang LM, Zhao J, Yu Z Uniform deposition of ultrathin polymer films on the surfaces of $Al_2O_3$ nanoparticles by a plasma treatment. Applied Physics Letters. 2010;**78**:1243-1245. DOI: http://dx.doi.org/10.1063/1.1352700

[57] Shi D, Lian J, He P, Wang LM, Xiao F, Yang L, Schulz MJ, Mast DB. Plasma coating of carbon nanofibers for enhanced dispersion and interfacial bonding in polymer composites. Applied Physics Letters. 2003;**83**:5301-5303. DOI: http://dx.doi.org/10.1063/1.1636521

[58] Schneiderman DK, Hillmyer MA. 50th anniversary perspective: There is a great future in sustainable polymers. Macromolecules. 2017;**50**:3733–3749. DOI: 10.1021/acs.macromol.7b00293

[59] Sezen-Edmonds M, Loo Y. Beyond doping and charge balancing: How polymer acid templates impact the properties of conducting polymer complexes. Journal of Physical Chemistry Letters. 2017;**8**:4530-4539. DOI: 10.1021/acs.jpclett.7b01785

[60] Zhang Y, Yuan J, Sun J, Ding G, Han L, Ling X, Ma W. Alkenyl carboxylic acid: Engineering the nanomorphology in polymer–polymer solar cells as solvent additive. ACS Applied Materials & Interfaces. 2017;**9**:13396-13405. DOI: 10.1021/acsami.7b02075

[61] Shi P, Amb CHA, Dyer AL, Reynolds JR. Fast switching water processable electrochromic polymers. ACS Applied Materials and Interfaces. 2012;**4**:6512-6521. DOI: 10.1021/am3015394

[62] Mohammadifar E, Bodaghi A, Dadkhahtehrani A, Nemati Kharat A, Adeli M, Haag R. Green synthesis of hyperbranched polyglycerol at room temperature. ACS Macro Letters. 2017;**6**:35-40. DOI: 10.1021/acsmacrolett.6b00804

[63] Laaziz SA, Raji M, Hilali E, Essabir H, Rodrigue D, Bouhfid R, Qaiss A. Bio-composites based on polylactic acid and argan nut shell: Production and properties. International Journal of Biological Macromolecules. 2017;**104**:30-42. DOI: 10.1016/j.ijbiomac.2017.05.184

[64] Laadila MA, Hegde K, Rouissi T, Brar SK, Galvez R, Sorelli L, Cheikh RB, Paiva M, Abokitse K. Green synthesis of novel biocomposites from treated cellulosic fibers and recycled bio-plastic polylactic acid. Journal of Cleaner Production. 2017;**164**:575-586. DOI: http://dx.doi.org/10.1016/j.jclepro.2017.06.235

[65] Jaso V, Glenn G, Klamczynski A, Petrovic ZS. Biodegradability study of polylactic acid/ thermoplastic polyurethane blends. Polymer Testing. 2015;**47**:1-3. DOI: 10.1016/j.polymertesting.2015.07.011

[66] Robin C, Cédric Lorthioir C, Amiel C, Fall A, Ovarlez G, Le Coeur C. Unexpected rheological behavior of concentrated poly(methacrylic acid) aqueous solutions. Macromolecules 2017;**50**:700-710. DOI: 10.1021/acs.macromol.6b01552

[67] Wu Y, Nizam NM, Ding X, Xu FJ. Rational design of peptide-functionalized poly (methacrylic acid). ACS Biomaterials Science & Engineering. 2017. DOI: 10.1021/acsbiomaterials.7b00584

Chapter 4

# The Excretion of Di- and Tricarboxylic Acids by Roots of Higher Plants Can Strongly Improve the Acquisition of Phosphate (P) by Plants in P-Fixing Soils

Jörg Gerke

Additional information is available at the end of the chapter

http://dx.doi.org/10.5772/intechopen.73033

**Abstract**

The excretion of di- and tricarboxylates by roots of higher plants represents a very efficient way to acquire phosphate (P) from soils, which are low in available P. Despite the extensive experimental work in evaluating the effect of carboxylates on the acquisition of soil phosphate in the fields of plant physiology, plant nutrition and soil chemistry in the last three decades, the effect of root excreted carboxylates on soil P acquisition by higher plants is still a matter of debate. This still ongoing debate has its origin in methodological deficits in experiments and misconceptions considering the role of carboxylates in P-fixing soils on P acquisition. The main, often found misconception is to assume that the parameter carboxylates in the rhizosphere soil solution is the most important parameter driving the P mobilization by carboxylates. Carboxylates in the soil solution are an easily degradable source of rhizosphere microorganisms, and carboxylates will induce P mobilization if the carboxylate is adsorbed onto the soil solid phase, a step which also will reduce microbial carboxylate degradation. Thus, often realistic concepts to quantify the effect of carboxylates on P mobilization are still missing. Second, the very important parameter of carboxylate accumulation at the rhizosphere soil solid is often ignored or is measured by weak extractants, which do not allow the quantitative determination of carboxylates bound to the soil solid phase. Both shortcomings are critically discussed in this chapter, and a experimental and mathematical procedure is presented to evaluate the effect of carboxylate excretion on P acquisition by higher plants.

**Keywords:** di- and tricarboxylic acids, adsorbed soil phosphate, humic-P complexes, P mobilization, acquisition of mobilized P, model calculations, strategies of P acquisition, rhizosphere of higher plants

## 1. Introduction

Since the publication of the paper of Dinkelaker et al. [1], increasing attention has been paid to the release of carboxylates by roots of higher plants and to its function in phosphate (P) mobilization in soil and P acquisition in P-fixing soils. Dinkelaker et al. [1] found in the rhizosphere of white lupin (*Lupinus albus* L.) grown in a calcareous soil, a high accumulation of macroscopic visible Ca-citrate precipitates as a result of selective citrate excretion by cluster roots of P-deficient white lupin plants.

From the viewpoint of soil chemistry, it has been relatively shown earlier that di- and tricarboxylates such as citrate, malate, oxalate or oxaloacetate can mobilize/dissolve strongly bound P in soil, which may contribute to the acquisition of P in P-deficient soils (Earl et al. [2]; Lopez- Hernandez et al. [3]).

Among the macronutrients of higher plants, N, K, P, Ca, Mg, S and P, phosphate is the nutrient which is strongly bound to the soil solid phase in many agricultural soils.

Worldwide, P deficiency of cultivated plants is a main, probably the main factor limiting plant yield [4–6]. The reserves for P fertilizer production are restricted compared to the reserves of other macronutrients [6].

In this chapter, the author considers the potential of the release of carboxylates for the improvement of the acquisition of P in P-fixing soils beginning with the potential of carboxylates excretion by the roots as a result of P deficiency, increasing the complexity of influencing parameters with each chapter of this contribution and finally, considering the acquisition of mobilized P by higher plants using mathematical modeling.

It will be shown that the adaptive excretion of carboxylates, especially citrate, and, to a lesser extent, oxalate is a very efficient mechanism to improve P acquisition by higher plants grown in P-fixing soils.

### 1.1. Adaptions of higher plants to P deficiency

Plants have developed several adaption mechanisms to P deficiency, morphological and physiological adaptions.

Morphological adaptions include an increasing root:shoot ratio with increasing P deficiency [7, 8], a higher number and length of root hairs [9, 10]. Both adaptions induce the same result, and the root surface is increased in relation to the shoot biomass, which may increase the acquisition of P from soils that are low in available P.

Physiological adaptions to P deficiency include, among others, the release of root exudates by P-starving plants, $H^+$ or $OH^-$ equivalents (see for a review Hinsinger [11]), reducing equivalents [12, 13] or the release of di- or tricarboxylic acid anions.

Plaxton and Tran [14] have recently provided an overview on the physiological adaptions of higher plants to P deficiency. In our context, the adaptions leading to an increased carboxylate efflux are of central importance.

An increased excretion of carboxylates during P deficiency requires two different simultaneous physiological adaptions, which are as follows: (1) an increase in the synthesis of carboxylates and (2) an increased rate of excretion (efflux) of the carboxylates.

a. An enhanced synthesis of carboxylate anions within the plants is correlated with the upregulation of the phosphoenol pyruvate carboxylase (PEPC), leading to an increased synthesis of oxaloacetate, an increased activity of malate dehydrogenase, leading to an increased reduction of oxaloacetate to malate, an increased activity of citrate synthase leading to an increased synthesis of citrate [14].

b. Organic acid anions are dissociated at the cytoplasmic pH within the roots. It was shown by Yan et al. [15] and Zhu et al. [16] that $H^+$- pumping ATPases in the plasma membrane of the roots are involved in an increased carboxylate efflux. At present, two separate transport processes for the release of carboxylates into the rhizosphere soil are assumed, an active $H^+$-efflux-driven carboxylate excretion process and a passive efflux of organic anions [14, 15].

### 1.2. Soil P forms, P availability and the interaction of soil P with carboxylates

Among the macronutrients, phosphorus, which is present in soil and plants nearly exclusively as orthophosphate anion (P) and its esters with organic alcohols, has, by far, the strongest affinity to the soil solid phase and consequently, the lowest solubility in the soil solution. In soil chemistry, the term buffering or buffer power is used to describe the intensity of bonding to the soil solid phase, which is described by the adsorption or desorption curve. High buffering or buffer power indicates that most of the added nutrient is bound to the solid phase and that the equilibrium soil solution concentration is low. Soils with high P buffering are defined as P-fixing soils. Such soils may contain high quantities of P; however, because of high concentration of P-sorbing sites, the P equilibrium soil solution concentration is low or even very low, sometimes near zero. The concentration or precisely the activity of P in the soil solution is important for the P acquisition of higher plants from soil because P is transported to the roots almost exclusively within the soil solution via diffusion [17, 18]. The diffusive flux strongly depends on the initial P concentration of the soil solution and consequently, the P concentration gradient within the rhizosphere soil solution (see for details [17–19]. Often, for maximum yield, higher plants need P soil solution concentrations above 1–5 μM P depending on the rooting density, the formation of root hairs, the root:shoot ratio and the formation of mycorrhiza. However, at very low P soil solution concentrations, an increase in the P absorbing root surface does not promote the acquisition of P adequately.

Under these conditions, strategies to increase the rhizosphere soil solution concentration of P by root exudates (chemical P mobilization) are promising and the central way to acquire P from strong P.

### 1.3. Chemical P mobilization in P-fixing soils

Several types of root exudates may increase the P solubility in soil and consequently increase the acquisition of P by the roots of higher plants.

As an adaption to P deficiency, plants can increase the efflux of protons followed by the acidification of the rhizosphere [11], they can increase the release of reducing agents (e.g., Tomasi et al. [13]), and they can increase the efflux of di- and tricarboxylates (see for a review, Gerke [17]).

The release of carboxylates is by far the most important way to acquire P even in P-fixing soils [17, 20].

To support this statement, information on the carboxylate efflux of different plant species and genotypes, the accumulation of carboxylates in the rhizosphere, the effect of carboxylates on the mobilization/dissolution of soil P and the acquisition of mobilized P by higher plants are required.

### 1.4. Carboxylate efflux by different plant species as affected by the P status

Proton release and carboxylate release are separate mechanisms in higher plants [15, 16]. However, often carboxylate and proton efflux is a coupled reaction to P deficiency.

As a result of P deficiency, many plant species show an increased carboxylate efflux, for example, white lupin [1, 21–23], alfalfa [24, 25], spinach [24, 26], chickpea [23]; red clover [24], yellow lupin [21, 27], radish buckwheat [28] and many members of the Proteaceae [29, 30] and of the Cyperaceae [31].

Many graminaceous species are considered to be ineffective in carboxylate excretion, but some graminaceous species show an increased carboxylate efflux to avoid Al- toxicity [32–35].

In buckwheat, oxalate efflux was increased during Al- toxicity [36].

Whether and to which extent Al- toxicity induced carboxylate excretion is controlled by reduced P acquisition or affects P acquisition is unknown.

The release of carboxylates by dicotyledonous plant species is not uniform along the root but is concentrated on the region behind the root tips [24, 26, 37–39].

The carboxylate release strongly depends on the P status of the plants and may be by a factor of 10–100 higher in P-deficient plants compared to high P plants.

Keerthisinghe et al. [40] found a very high citrate release in fully developed cluster roots of white lupin with a maximum efflux of about 6.1 [nmol $h^{-1}$ $cm^{-1}$], decreasing in younger and old cluster root segments by a factor of 10 or even more.

Neumann et al. [41] confirmed these results; however, Watts and Evans (1999) measured a maximum citrate in cluster roots of white lupin of 33 [nmol $min^{-1}$ $m^{-1}$], which is about three times higher than the values reported by Keerthisinghe et al. [40].

Yellow lupin, a plant species that also forms cluster roots [21, 27], exhibits a maximum citrate efflux within the cluster roots of about 70–80% than that of the cluster roots of white lupin [27].

Cluster roots of the Proteaceae, for example, Hakea prostrata shows a carboxylate efflux of cluster roots, which is much higher than that of the white lupin [31, 42]. In Hakea undulate, malate is the main carboxylate, which is excreted with a root efflux being two times higher

http://dx.doi.org/10.5772/intechopen.73033

than the citrate efflux in white lupin [43]. Members of the Cyperaceae form "dauciform" roots as an adaption to P deficiency, which exhibit carboxylate efflux rates which are similar to the cluster roots of Proteaceae species [31].

Also, plant species that do not form cluster or dauciform roots can show a high carboxylate efflux during P deficiency. The region of high carboxylate efflux is often restricted to about 1–2 mm behind the root tips. For example, Beißner [37] showed for sugar beet, an increased oxalate efflux during P deficiency, the carboxylate efflux being by a factor of 3–4 higher, 1–2 mm behind the root tips, compared to the overall efflux of the whole root system.

Gerke [24] investigated the carboxylate efflux of several plant species, grown in quartz sand at four levels of P supply (**Figures 1–6**).

As the shoot P concentrations decreased, the efflux of carboxylates increased in all plant species was investigated (**Figures 1–6**).

However, the main carboxylates differed between the plant species. In legumes such as white clover, red clover or alfalfa, citrate is the dominant anion, whereas in spinach, oxalate is dominant and in the Brassicaceae species, Chinese cabbage (*Brassica Chinensis*), malate is dominant. Also, in rape, malate is the dominant anion [38].

In graminaceous species such as rye grass, the carboxylate efflux is extremely low, even at strong P deficiency (**Figure 6**), whereas especially in alfalfa and red clover, the overall citrate efflux was very high under P deficiency (**Figures 1** and **2**).

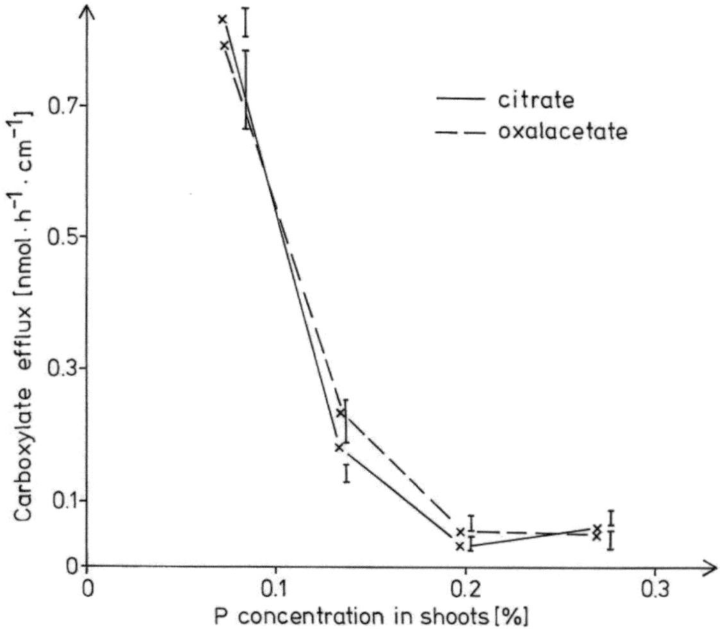


**Figure 1.** Carboxylate efflux by alfalfa (*Medicago sativa* L.) as affected by shoot P concentrations. (Modified from Gerke [24]).

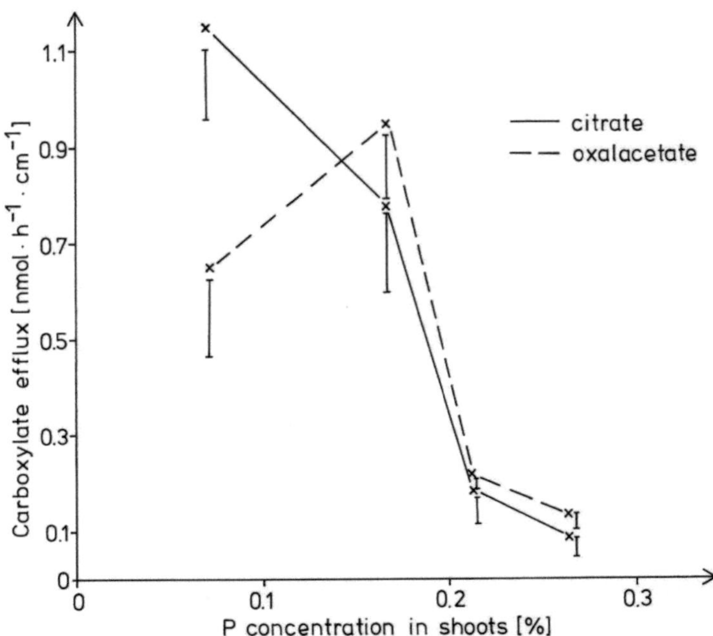


**Figure 2.** Carboxylate efflux by red clover (*Trifolium pratense* L.) as affected by shoot P concentrations. (Modified from Gerke [24]).

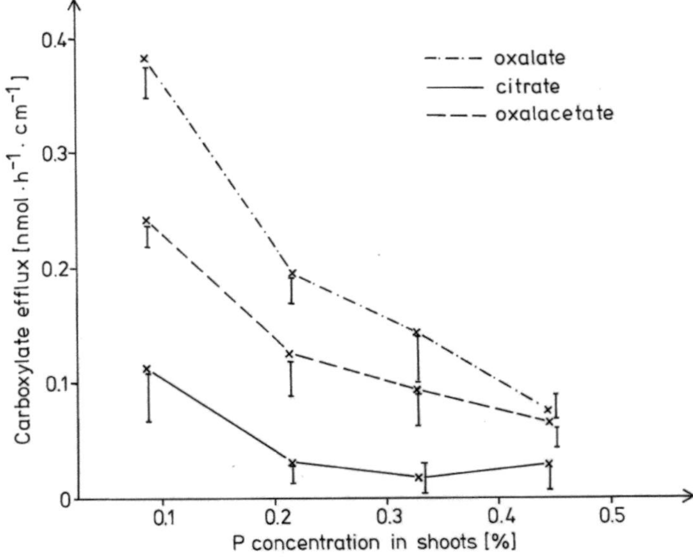


**Figure 3.** Carboxylate efflux by spinach (*Spinacia oleracea* L.) as affected by shoot P concentrations. (Modified from Gerke [24]).

## 1.5. Accumulation of carboxylates in the rhizosphere

A high or even very high carboxylate efflux by roots of P-deficient plants is no guarantee for its accumulation within the rhizosphere soil. Carboxylates which are excreted into the soil solution

http://dx.doi.org/10.5772/intechopen.73033

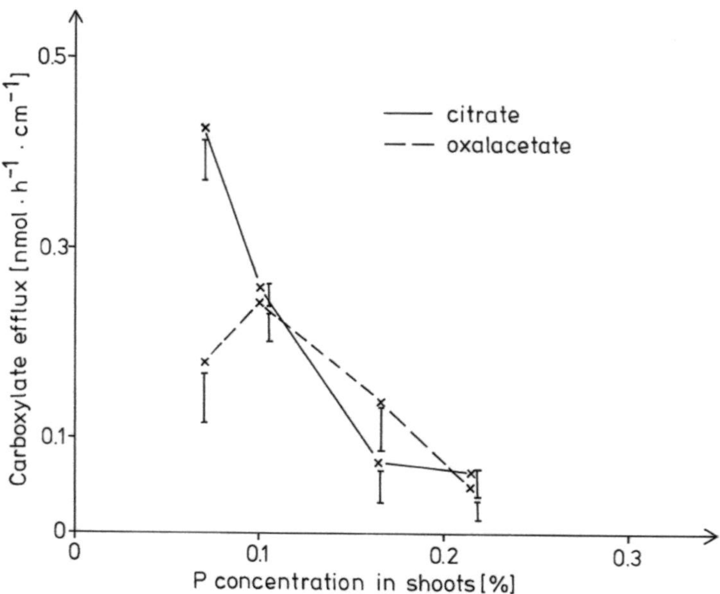


**Figure 4.** Carboxylate efflux by white clover as affected by shoot P concentrations. (Modified from Gerke [24]).

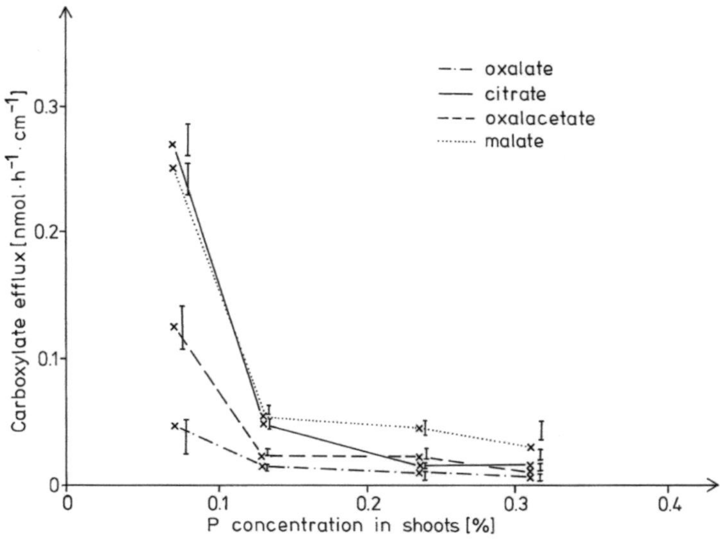


**Figure 5.** Carboxylate efflux by Chinese cabbage as affected by shoot P concentrations. (Modified from Gerke [24]).

are easily degradable C- source for soil microorganisms, which sometimes has been taken as an argument for the questioning of the relevance of carboxylates for P acquisition in soil (see e.g., the review of Richardson et al. [44]). However, the same review group of Richardson et al. [44] simplified the role of carboxylates in the rhizosphere to the question of carboxylate concentration in the soil solution assuming the soil solution being the main reservoir for carboxylates. However, the interaction of the excreted carboxylates and the soil solid phase is decisive for its effect on soil P solubility and the P acquisition by the plants in P-fixing soils.

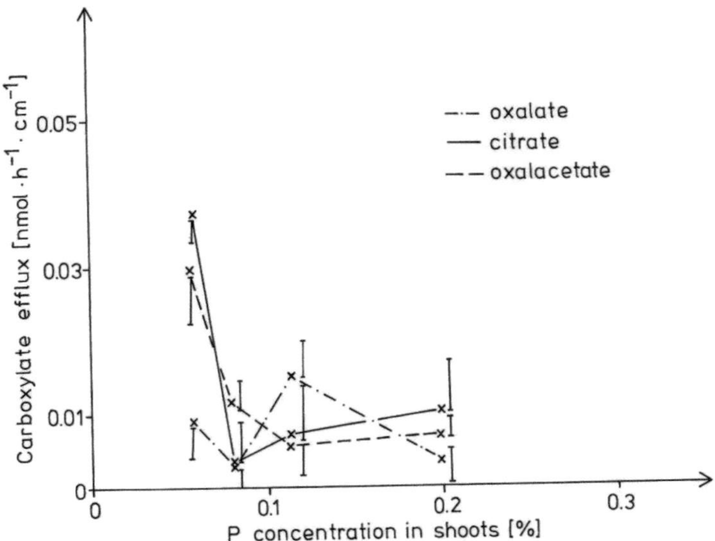


**Figure 6.** Carboxylate efflux by ryegrass (*Lolium perenne* L.) as affected by shoot P concentrations. (Modified from Gerke [24]).

The carboxylate concentration in the soil solution depends on the carboxylate excretion and the buffering of the carboxylates in soil and is not very relevant. Relevant for P mobilization is the concentration of carboxylates at the soil solid phase. The sorbed carboxylates can desorb P by occupying the P-sorption sites, and they can induce dissolution processes of P-sorbing surfaces such as Al (Fe)-(hydr)oxides or humic-Al(Fe) complexes [45, 46].

The result of these reactions between carboxylates and the soil solid phase is an increase in the P soil solution concentrations as well as an increase in Fe and Al solubility, which was experimentally shown for cluster root rhizosphere soil solution of white lupin [47] and is often reported in model experiments (e.g., Fox et al. [48]; Takeda et al. [49]; Gerke [46]).

Di- and tricarboxylates in the soil solution are easily degraded by microorganisms. If the excreted carboxylate is bound to the rhizosphere soil solid by a fast adsorption reaction, carboxylate degradation is strongly retarded.

Boudot [50] and Jones and Edwards [51] showed that the sorption of carboxylates to alumosilicates, Al-oxides and Fe- oxides strongly prevented microbial degradation. Boudot [50] investigated the effect of $^{14}C$-citrate adsorption to Al containing soil minerals. Free citrate was mineralized to about 70–80% within 10 days. At a high Al/citrate ratio, adsorption of citrate decreased the mineralization of citrate to zero. Similar results were found by Jones and Edwards [51] in the system citrate/Fe-oxides. Boudot [50] also showed that the formation of Al-citrate complexes in the solution strongly reduced citrate mineralization.

It has sometimes been speculated that acidification of the rhizosphere, for example, by legumes may reduce carboxylate mineralization within the rhizosphere soil (e.g., Lambers et al. [52]). At present, the effect of acidification on carboxylate adsorption and its consequent effect on P mobilization are estimated to be more important (Gerke, 2000a).

The accumulation of carboxylates in the rhizosphere is remarkable. Dinkelaker et al. [1] found more than 50 μmol citrate/g soil in the cluster root rhizosphere of white lupin. Gerke [17, 24] found between 12 and 88 μmol citrate/g soil in the cluster root rhizosphere of white lupin grown in different soils at different P levels depending on the method of citrate determination.

Often, carboxylates in the rhizosphere soil of different plant species are quantified after extraction with mild extractants, such as distilled water [53, 54], dilute salt solutions [23, 55, 56] or dilute acid [57], leading to a low recovery of the carboxylates. Gerke et al. [47] found citrate concentrations in the cluster root rhizosphere of between 66 and 88 [μmol citrate/g soil] determined by direct infrared spectroscopy (DRIFT spectroscopy), whereas the quantity of citrate extracted with water was below the detection limit.

Cluster roots are a peculiar adaption of relatively few plant species, among them, both white lupin and yellow lupin are cultivated plant species.

In the noncluster root forming plant species, red clover, Gerke and Meyer [58] showed a citrate accumulation of more than 12 [μmol citrate/g soil] in close proximity to the roots of P-deficient plants.

### 1.6. The effect of carboxylates on the mobilization/dissolution of soil phosphate

Several carboxylates can strongly increase the P soil solution concentration.

This was relatively shown earlier, for example, by Earl et al. [2] and Lopez-Hernandez et al. [3].

More detailed results were reported by Gerke [24] and Gerke et al. [59].

Often, with increasing carboxylate application and accumulation at the soil solid phase, the P concentrations in the soil solution increase exponentially (**Figures 7** and **8**) [59].

From **Figures 7** and **8** it can be seen that changes in soil pH alone have only a minor impact on the soil P solubility but pH changes strongly affect the effectiveness of citrate to mobilize soil P. The exponential relation can also be shown, for example, for oxalate [24, 37].

**Table 1** shows that citrate and, to some extent, oxalate are very efficient in mobilizing soil P, whereas malate and oxaloacetate are relatively inefficient.

### 1.7. The acquisition of mobilized P by higher plants

Many experimental results demonstrate that P deficiency in higher plants increases the carboxylate efflux. Also, many experimental results show that citrate, oxalate, to some extent, malate and other carboxylates can mobilize P in P-fixing soils.

The separate views on the physiology of carboxylate excretion and root morphology or, on the other hand, the rhizosphere chemistry of carboxylates and its impact on the P solubility in soil does not prove the relevance of carboxylate excretion on the P acquisition of higher plants.

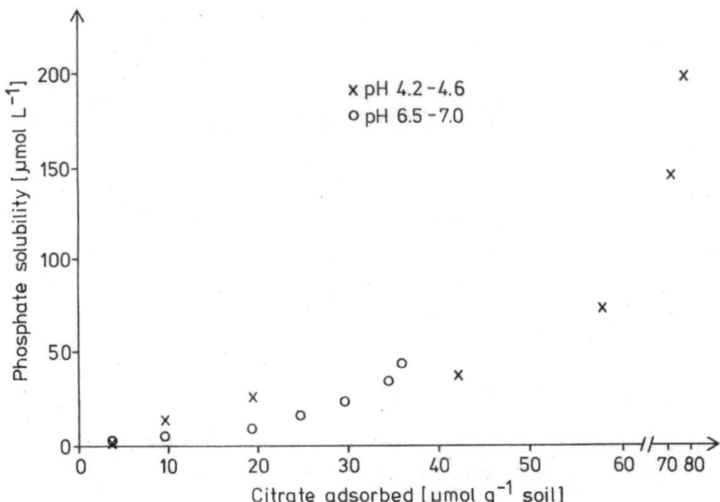


**Figure 7.** Relation between the citrate concentration in soil and the P solubility at different pH in a Ferralsol. (Modified from Gerke [24]).

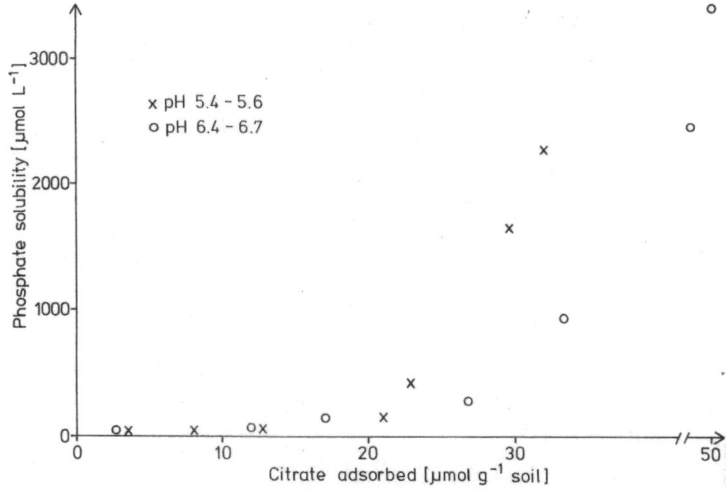


**Figure 8.** Relation between the citrate concentration in soil and the P solubility at different pH in a humic Podzol. (Modified from Gerke [24]).

The combination of both views, including carboxylate efflux, the accumulation of carboxylates in the rhizosphere soil, the chemistry of P mobilization and the uptake of mobilized P by the roots may help to evaluate the contribution of carboxylate excretion to P acquisition by higher plants.

For this purpose, mathematical models are a useful tool. Also, such models allow, by the aid of sensitivity analysis, to evaluate the contribution of different parameters to the acquisition of P by higher plants.

| | | Phosphate | Fe + Al (μmol/l) |
|---|---|---|---|
| **Luvisol** | Citrate | 85.6 | 2340.5 |
| pH 7.7–8.1 | Oxalate | 70.4 | 1234.3 |
| | Malate | 36.4 | 812.3 |
| | Oxaloacetate | 32.2 | 972.4 |
| **Ferralsol** | Citrate | 198.5 | 9420.1 |
| pH 4.1–4.7 | Oxalate | 28.3 | 1314.4 |
| | Malate | 25.2 | 2810.4 |
| | Oxaloacetate | 7.5 | 71.5 |

**Table 1.** Maximum phosphate and iron + aluminum mobilization by organic anions.

Nye [60, 61] developed a mathematical model to quantitatively describe the influx of P as affected by the efflux of mobilizing agents.

Gerke [24] was the first who used the mathematical framework of Nye [60, 61] to quantify the influx of mobilized P as a result of a definitive carboxylate efflux. Some of the results of the calculations are shown in **Figures 9** and **10**.

For the three plant species such as rye grass, white clover and red clover, the maximum citrate efflux was used in the calculations (from **Figures 2, 4,** and **6**). For the accumulation of citrate in soil, we assumed two strongly differing values in order to include a possible range of accumulation (see in detail Gerke [24]).

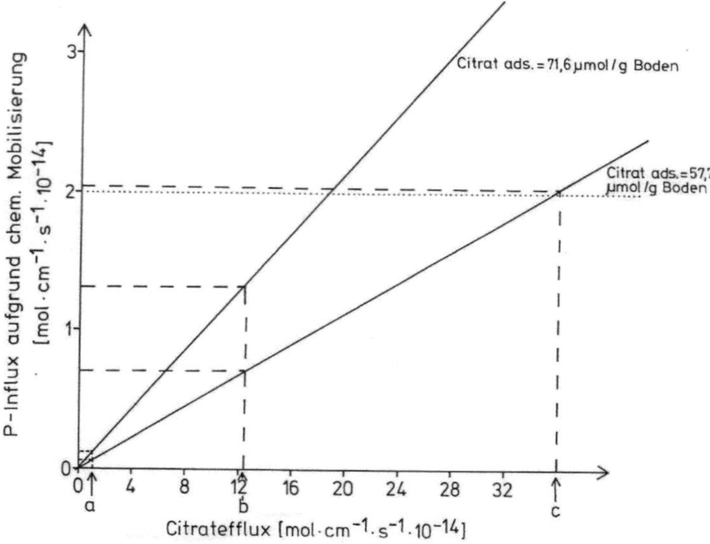


**Figure 9.** Influx of mobilized P as related to maximum citrate efflux of plants grown in a Ferralsol by (a) rye grass, (b) white clover and (c) red clover. (Modified from Gerke [24]).

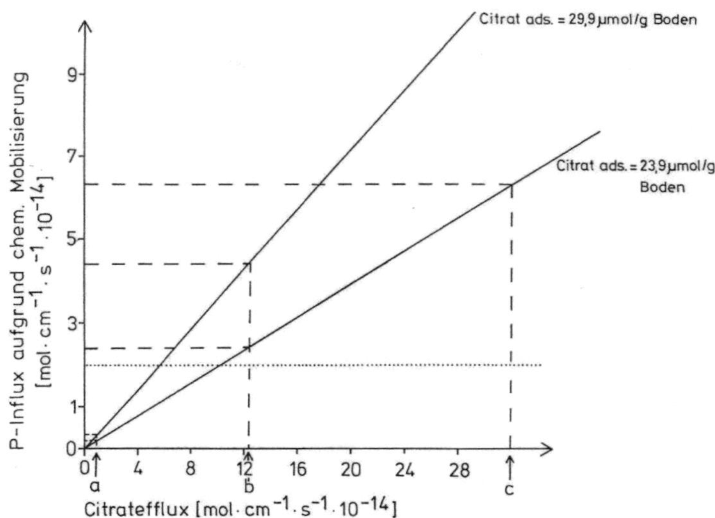


**Figure 10.** Influx of mobilized P as related to citrate efflux of plants grown in a humic Podzol by (a) rye grass, (b) white clover and (c) red clover. (Modified from Gerke [24]).

In none of the soils at either citrate accumulation level did the maximum citrate efflux of rye grass have any effect on the acquisition of P. The high overall citrate efflux of P-deficient red clover strongly increased the influx of mobilized P over the required level in all cases. White clover with a medium citrate efflux strongly increased the P influx of mobilized P in a soil with humic-Al(Fe)-P complexes as dominant P fraction. Compared to soils with mainly Al(Fe)-oxide-P complexes, humic-associated P is much easier to mobilize by carboxylates [46, 58].

The results shown in **Figures 9** and **10** show that citrate excretion by P-deficient roots is an exceptional efficient way to improve P acquisition in P-fixing soils.

In their review on P efficiency, Richardson et al. [44] presented a separate chapter titled: "Can the release of organic anions from roots mobilize phosphorus?"

In this chapter, Richardson et al. [44] related [5–50 µmol citrate/g rhizosphere soil] to 1–10 [mM] citrate in the soil solution and then stated that 1 [mM] citrate or more is effective for P mobilization in some soils. This view on the rhizosphere chemistry of carboxylates and phosphate is misleading for at least two reasons.

First, the buffering of carboxylates in soil is a strongly varying soil parameter and cannot be generalized to a relation of 1–10 [mM] citrate versus 5–50 [µmol citrate/g soil], suggesting this as the general range of citrate buffering in all soils. Gerke [17, 24] showed for a strongly anion-fixing Ferralsol, that citrate concentration in the soil solution of less than 10 µM corresponds to more than 60 [µmol citrate/g soil] at the soil solid phase suggesting a much wider range of citrate buffering.

Second, it is a misleading view that carboxylates in the soil solution are responsible for the mobilization of P. The initial or essential step in P mobilization is the adsorption of the carboxylates to the soil solid phase where the carboxylate-mediated P desorption or dissolution of P-sorbing surfaces is initiated. In this context, simultaneous acidification and carboxylate excretion may increase the dissolution of P-sorbing sites more efficient than carboxylate excretion alone [21].

Finally, the role of root-derived carboxylates on the acquisition of P esters should be considered. This topic has been reviewed by Gerke [20]. In general, phosphate monoesters may accumulate in soils among which higher phosphorylated inositol phosphates account for the dominant proportion of ester-P in many soils [20].

Worldwide, more than 51 millions of tons of phytate (myoinositol hexakisphosphate) are annually produced in crops, seeds and fruits, which is equivalent to about 66% of the P annually applied to agricultural land [62]. Higher phosphorylated inositol phosphates are strongly bound to the soil solid phase often much stronger than the orthophosphate anion [20, 63]. The initial step of P acquisition from inositol phosphates is its mobilization.

Adams and Pate [64] showed that in soil, white lupin but not narrow leaf lupin can acquire P from phytate. White lupin forms root clusters with extensive citrate excretion, but narrow leaf lupin forms no root clusters [27].

The mobilization by carboxylates is the rate limiting step in acquisition of P from phytate and similar molecules and not the reduced activity of phytases. For a detailed discussion, see Gerke [20].

## 2. Conclusions

The partly contrasting results concerning the role of root-released carboxylates on P acquisition by higher plants are mainly due to deficits in the concepts of research and experimental methods.

Carboxylates will affect the soil P solubility after adsorption to the soil solid phase and not dissolved in the soil solution. It is a misconception to consider exclusively the carboxylate concentration in the rhizosphere soil solution and not the carboxylate concentration in the soil solid phase.

The determination of carboxylate bound to the soil solid phase is often ignored or is determined by deficient methods. Often, weak extractants such as dilute acids or dilute salt solutions are used, which extract a very low proportion of rhizosphere soil-bound carboxylates.

Considering the range of carboxylate efflux of P-deficient plants, especially cluster root forming plants such as white and yellow lupin, and species such as red clover, alfalfa, sugar beet and spinach may strongly mobilize the soil P and acquire the mobilized P.

Citrate and, to some extent, oxalate are the anions, which are most efficient with respect to P mobilization.

The most important organic P forms in soil, higher phosphorylated inositol phosphates are strongly bound to the soil solid phase and can be mobilized by carboxylates similar to the orthophosphate anion.

The quantitative effect of carboxylate excretion on the acquisition of soil P by higher plants requires an integrative concept, including all relevant soil and plant parameters. Such a framework may be a mathematical model based on experimentally determined parameters.

## Author details

Jörg Gerke

Address all correspondence to: gerke.rukieten@t-online.de

Reetwiesenhof, Rukieten, Germany

## References

[1] Dinkelaker B, Römheld V, Marschner H. Citric acid excretion and precipitation of calcium citrate in the rhizosphere of white lupin (*Lupinus albus* L.). Plant, Cell Environment. 1989;**12**:185-192

[2] Earl K, Syers JK, Mc Laughlin R. Origin of the effect of citrate, tartrate and acetate on phosphate sorption by soils and synthetic gels. Soil Science Society of American Journal. 1979;**43**:674-678

[3] Lopez- Hernandez D, Flores D, Siegert G, Rodriguez J. The effect of some organic anions on phosphate removal from acid and calcareous soils. Soil Science. 1986;**128**:321-326

[4] Kirkby AE, Johnston AE. Soil and fertilizer phosphorus in relation to crop nutrition. In: Hammond JP, White PI, editors. The Ecophysiology of Plant Phosphorus Interactions. New York: Springer; 2008. pp. 177-223

[5] Marschner P, editor. Marschners Mineral Nutrition of Higher Plants. Amsterdam: Elsevier; 2012

[6] Römer W. Ansätze für eine effizientere Nutzung des Phospors auf der Basis experimenteller Befunde. Berichte Uber Landwirtschaft. 2009;**87**:5-30

[7] Föhse D, Claassen N, Jungk A. Phosphorus efficiency of plants. I. External and internal P requirement and P uptake efficiency of different plant species. Plant Soil. 1988;**110**:101-109

[8] Lynch JP, Brown KM. Topsoil foraging- an architectural adaption of plants to low phosphorus availability. Plant Soil. 2001;**237**:225-237

[9] Föhse D, Jungk A. Influence of phosphate and nitrate supply on root hair formation of rape, spinach, and tomato plants. Plant Soil. 1983;**74**:359-368

[10] Föhse D, Claassen N, Jungk A. Phosphorus efficiency of plants. II. Significance of root radius, root hairs, and cation-anion balance for phosphorus influx in seven plant species. Plant Soil. 1991;**132**:261-272

[11] Hinsinger P. Bioavailability of soil inorganic P in the rhizosphere as affected by root-induced chemical changes: A review. Plant Soil. 2001;**237**:173-195

[12] Marscher H, Römheld V. Strategies of plants for acquisition of iron. Plant Soil. 1994; **165**:261-274

[13] Tomasi N, Weisskopf L, Renella G, Landi L, Pinton R, Varini Z, Nannipieri P, Torrent J, Martinoia E, Cesco S. Flavonoids of white lupin roots participate in phosphorus mobilization from soil. Soil Biology Biochemistry. 2008;**38**:1971-1974

[14] Plaxton WC, Tran HT. Metabolic adaptions of phosphate starved plants. Plant Physiology. 2011;**156**:1006-1015

[15] Yan F, Zhu Y, Müller C, Zörb C, Schubert S. Adaption of $H^+$- pumping and plasma membrane $H^+$ Atpase activity in proteoid roots of white lupin under phosphorus deficiency. Plant Physiology. 2002;**129**:50-63

[16] Zhu Y, Yan F, Zörb C, Schubert S. A link between citrate and proton release by proteoid roots of white lupin (*Lupinus albus L.*) grown under phosphorus-deficient conditions. Plant Cell Physiology. 2005;**46**:892-901

[17] Gerke J. The acquisition of phosphate by higher plants: Effect of carboxylate release by the roots. A critical review. Journal of Plant Nutrition and Soil Science. 2015;**178**:351-364

[18] Jungk A. Dynamics of nutrient movement at the soil-root interface. In: Waisel Y, Eshel A, Kafkafi U, editors. Plant Roots, the Hidden Half. New York: Basel Dekker; 2002. pp. 587-616

[19] Jungk A. Root hairs and the acquisition of plant nutrients from soil. Journal of Plant Nutrition and Soil Science. 2001;**164**:121-129

[20] Gerke J. Phytate (inositol hexakisphosphate) in soil and phosphate acquisition from inositol phosphates by higher plants. Plants. 2015;**4**:253-266

[21] Egle K, Römer W, Gerke J, Keller H. The influence of P nutrition on organic acid exudation in three lupin species. In: Van Santen E, Wink M, Weissmann S, Römer P, editors. Lupin: An Ancient Crop for a New Millenium. New Zealand: Canterbury; 1999. pp. 249-251

[22] Neumann G, Römheld V. Root excretion of carboxylic acids and protons in phosphorus-deficient plants. Plant Soil. 1999;**211**:121-130

[23] Veneklaas EJ, Stevens J, Cawthray GR, Turner S, Grigg AM, Lambers H. Chickpea and white lupin carboxylates vary with soil properties and enhance phosphorus uptake. Plant Soil. 2003;**248**:187-197

[24] Gerke J. Chemische Prozesse der Nährstoffmobilsierung in der Rhizosphäre und ihre Bedeutung für den Übergang vom Boden in die Pflanze. Göttingen: Cuvillier- Verlag; 1995

[25] Lipton D, Blanchar R, Blevins D. Citrate, malate and succinate concentrations in exudates from P-deficient and P-stressed *Medicago sativa L.* seedlings. Plant Physiology. 1987;**85**:315-317

[26] Keller H. Einfluss wurzelbürtiger organischer Säuren auf das Cu-,Zn- und Cd- Aneignungsvermögenvon Spinatgenotypen [PhD thesis]. Germany: University of Kaiserslautern; 2000

[27] Hocking P, Jeffery S. Cluster root production and organic anion exudation in a group of old-world lupins and a new world lupin. Plant Soil. 2004;**258**:135-150

[28] Possinger AR, Byrne LB, Breen NE. Effect of buckwheat (*Fagopyrum esculentum*) on soil phosphorus availability and organic acids. Journal of Plant Nutrition and Soil Science. 2013;**176**:16-18

[29] Dinkelaker B, Hengeler C, Marschner H. Distribution and function of proteoid roots and other root clusters. Botanica Acta. 1995;**108**:169-276

[30] Roelofs RF, Rengel Z, Cawthray GR, Dixon KW, Lambers H. Exudation of carboxylates in Australian Proteaceae: Chemical composition. Plant Cell Environment. 2001;**24**:891-904

[31] Shane MW, Cawthray GR, Cramer MD, Kuo J, Lambers H. Specialized dauciform roots of Cyperaceae are structurally distinct but functionally analogous with cluster roots. Plant Cell Environment. 2006;**29**:1989-1999

[32] Delhaize E, Ryan PR, Randall PJ. Aluminum tolerance in wheat (Triticum aestivum L.). II. Aluminum-stimulated excretion of malic acid from root apices. Plant Physiology. 1993;**103**:695-702

[33] Li XF, Ma FM, Matsumoto H. Aluminum- induced secretion of both, citrate and malate in rye. Plant Soil. 2002;**242**:235-243

[34] Stass A, Smit I, Eticha D, Oettler G, Horst WJ. The significance of organic anion exudation for the aluminum resistance of primary triticale derived from wheat and rye parents differing in aluminum resistance. Journal of Plant Nutrition and Soil Science. 2008;**171**:634-642

[35] Zhao Z, Ma F, Sato K, Takeda K. Differential Al resistance and citrate secretion in barley (*Hordeum vulgare L.*). Planta. 2003;**217**:794-800

[36] Ma Z, Miasaka SC. Oxalate exudation as a mechanism of aluminum tolerance in taro. Plant Physiology. 1998;**118**:861-865

[37] Beißner L. Mobilisierung von Phosphor aus organischen und anorganischen Verbindungen durch Zuckerrübenwurzeln [PhD thesis]. Germany: Georg-August Universität Göttingen; 1997

[38] Hoffland E, Findenegg G, Nelemans J. Solubilization of rock phosphate. II. Local exudation of organic acids as a response to P starvation. Plant Soil. 1989;**113**:161-165

[39] Römer W, Kang DG, Egle K, Gerke J, Keller H. The acquisition of cadmium by *Lupinus albus L., Lupinus angustifolius L.* and *Lolium multiflorum Lam.* Journal of Plant Nutrition and Soil Science. 2000;**163**:623-632

[40] Keerthisinghe G, Hocking P, Ryan PR, Delhaize E. Proteoid roots of lupin (*Lupinus albus L.*): Effects of phosphorus supply on formation and spatial variation in citrate efflux and enzyme activity. Plant Cell Environment. 1998;**21**:467-478

[41] Neumann G, Massoneau A, Martinoia E, Römheld V. Phsiological adaptions to phosphorus deficiency during proteoid root development in white lupin. Planta. 1999;**208**:373-382

[42] Shane MW, Cramer MD, Funayama-Nogushi S, Cawthray GR, Millar AH, Day DA, Lambers H. Developmental physiology of cluster-root carboxylate synthesis and exudation in harsh Hakea. Expression of phosphoenolpyruvate carboxylates and alternative oxidase. Plant Physiology. 2004;**135**:549-560

[43] Dinkelaker B, Hengeler C, Neumann G, Eltrop L, Marschner H. Root exudates and mobilization of nutrients. In: Renneberg H, Escherich W, Ziegler H, editors. Trees-Contribution to Modern Tree Physiology. Leiden: Backhuys Publishers; 1997. pp. 441-452

[44] Richardson AE, Lynch JP, et al. Plant and microbial strategies to improve phosphorus efficiency of agriculture. Plant Soil. 2011;**349**:121-156

[45] Geelhood JS, van Riemsdijk WH, Fingenegg GR. Simulation of the effect of citrate exudation from roots on the plant availability of phosphate adsorbed on goethite. European Journal of Soil Science. 1999;**50**:379-390

[46] Gerke J. Humic (organic matter)-Al(Fe)- phosphate complexes: An underestimated phosphate form in soils and source of plant- available phosphate. Soil Science. 2010; **175**:417-425

[47] Gerke J, Römer W, Jungk A. The excretion of citric and malic acid by proteoid roots of *Lupinus albus* L.: Effects on soil solution concentrations of phosphate, iron and aluminum in the proteoid rhizosphere in samples of an oxisol and a luvisol. Journal of Plant Nutrition and Soil Science. 1994;**157**:289-294

[48] Fox T, Comerford N, Mc Fee W. Phosphorus and aluminum release from a spodic horizon mediated by organic acids. Soil Science Society of American Journal. 1990;**54**:1763-1767

[49] Takeda A, Tsukada H, Takaku Y, Hisamatsu S. Fractionation of metal complexes with dissolved organic matter in a rhizosphere soil solution of a humus- rich Andosol using size exclusion chromatography with inductively coupled plasma-mass spectrometry. Soil Science and Plant Nutrition. 2009;**55**:349-357

[50] Boudot JP. Relative efficiency of complexed aluminum, noncrystralline Al hydroxide, allophane and imogolite in retarding biodegradation of citric acid. Geoderma. 1992;**52**:29-39

[51] Jones DL, Edwards AC. Influence of sorption on the biological utilization of two simple carbon substrates. Soil Biology and Biochemistry. 1998;**30**:1895-1902

[52] Lambers H, Clements JC, Nelson MN. How a phosphorus- acquisition strategy based on carboxylate exudation powers the success and agronomic potential of lupines (Lupinus, Fabaceae). American Journal of Botany. 2013;**100**:263-288

[53] Li HG, Shen JB, Zhang FS, Lambers H. Localized application of soil organic matter shifts distribution of cluster roots of white lupin in the soil profile due to localized release of phosphorus. Annals of Botany. 2010;**105**:585-593

[54] Shen J, Rengel Z, Tang C, Zhang F. Role of phosphorus nutrition in development of cluster roots and release of carboxylates in soil-grown *Lupinus albus*. Plant Soil. 2003;**248**:199-206

[55] Nuruzzaman M, Lambers H, Bolland MD, Veneklaas E. Distribution of carboxylates and acid phosphatase and depletion of different phosphorus fractions in the rhizosphere of a cereal and three grain legumes. Plant Soil. 2006;**281**:109-120

[56] Wouterloud M, Cawthray GR, Scalan T, Lambers H. Carboxylate concentrations in the rhizosphere of lateral roots of chickpea (*Cicer arientinum L.*) increase during plant development but are not correlated with phosphorus status in soil or plants. New Phytology. 2004;**162**:745-753

[57] Kirk GJ, Santos EE, Santos MB. Phosphate solubilization by organic acid excretion from rice growing in aerobic soil: Rates of excretion and decomposition, effects on rhizosphere pH and effects on phosphate solubility and uptake. New Phytology. 1999;**142**:185-200

[58] Gerke J, Meyer U. Phosphate acquisition by red clover and black mustard on a humic podzol. Journal of Plant Nutrition. 1995;**18**:2409-2429

[59] Gerke J, Beißner L, Römer W. The quantitative effect of chemical phosphate mobilization by carboxylate anions on P uptake by a single root. I. The basic concept and determination of soil parameters. Journal of Plant Nutrition and Soil Science. 2000;**163**:207-212

[60] Nye PH. The diffusion of two interacting solutes in soil. Journal of Soil Science. 1983; **34**:677-691

[61] Nye PH. On estimating the uptake of nutrients solubilzed near roots or other surfaces. Journal of Soil Science. 1984;**35**:439-446

[62] Mullaney EJ, Ullah AH. Phytases: Attributes, catalytic mechanisms and applications. In: Turner BL, Richardson AE, Mullaney EJ, editors. Inositol Phosphates: Linking Agriculture and the Environment. Cambridge: Wallingford; 2007. pp. 97-110

[63] Mc Kercher R, Anderson G. Organic phosphate sorption by neutral and basic soils. Communicaions in Soil Science and Plant Analysis. 1989;**20**:723-732

[64] Adams M, Pate JS. Availability of organic and inorganic forms of phosphorus to lupins. Plant Soil. 1992;**145**:197-113

Chapter 5

# Isolation and GC Analysis of Fatty Acids: Study Case of Stinging Nettle Leaves

Saša Đurović, Saša Šorgić, Saša Popov,
Marija Radojković and Zoran Zeković

Additional information is available at the end of the chapter

http://dx.doi.org/10.5772/intechopen.73533

**Abstract**

*Urtica dioica* L. is perennial plant with remarkable medical properties and has been widely used in folk medicine and as a food. Fatty acids presented in its leaves were extracted applying three different techniques: classic, automated Soxhlet, and supercritical fluid extraction (SFE). SFE was performed at three different pressures (100, 200, and 300 bar) and two different temperatures (40 and 60°C). Obtained extract was analyzed using GC-FID analytical technique in order to obtain fatty acid profile samples. The highest yield was obtained in the case of automated Soxhlet extraction (218.907 mg/g), while the lowest was observed in the case of classic extraction (15.031 mg/g). Soxhlet approach provided the highest yield of saturated fatty acids (173.348 mg/g), while supercritical fluid extraction gave better results in the case of unsaturated fatty acids (60.062 mg/g). Deeper analysis of SFE extracts revealed relationship between extraction parameters (temperature and pressure) and yields of fatty acids where lower pressure ensured higher yield of unsaturated while higher pressure gave higher yield of saturated acids. On the other hand, increasing in temperature at isobaric conditions negatively influenced the yield of fatty acids except in the case of 300 bar where yield was higher at 60°C.

**Keywords:** stinging nettle, leaves, extraction, fatty acids, GC-FID analysis

## 1. Introduction

*Urtica dioica* L. (stinging nettle) is perennial, wild-growing plant from Urticaceae botanical family [1]. This plant is generally known for biological activity and positive effect which expresses on human health. It has long history of application in folk medicine where it has

been used for treatment of flailing arthritis or paralytic limbs, stimulation of circulation, and warming the joints and extremities [2]. There are wide ranges of studies which have dealt with biological activity of this plant. They have proved antioxidant, antimicrobial, anti-inflammatory, antiulcer, and analgesic properties of plant and its extracts [2–7]. Besides application in medicine, this plant has also been used in human nutrition for a long time. It has been harvested commercially for high content of chlorophylls, which has been used as coloring agent (E140) in food and medical products [8].

Previously conducted studies showed that stinging nettle contains wide diversity of biologically active compounds. There are essential oils and terpenoids as the main components [9, 10], carotenoids [4, 9, 11], fatty acids [9, 11, 12], different phenolic and polyphenolic compounds [1, 3, 9, 13, 14], essential amino acids, chlorophyll, vitamin C, vitamin K, vitamins of B series, tannins, carbohydrates, sterols, polysaccharides, isolectins [4, 9, 11, 15], and minerals [9, 16, 17]. The most important carotenoids are β-carotene, neoxanthin, lycopene, and lutein [11]; fatty acids are palmitic and *cis*-9,12 linoleic and α-linolenic acids [9, 11, 12], while iron has been marked as the most important mineral [9].

Fatty acids or generally fats are presented in many different forms such as saturated, monounsaturated, unsaturated, omega-3, omega-6, trans, or partially hydrogenated fats. These terms are widely presented in media regarding the health subject. Fats represent the most efficient way of storing excess energy. Besides this role, fats represent building blocks of lipid membranes (in the form of phospholipids), play significant roles in maintaining proper function of brain, and participate in process of signal transduction (in the form of diacylglycerols). Certain fats such as linoleic and arachidonic are essential due to inability of human body to synthetize them. Thus, their presence is very important because their involvement in biosynthesis of eicosanoids. This class of biomolecules is further involved in wide range of processes in organism such as platelet aggregation, anti-inflammatory response, allergic reactions, etc. Polyunsaturated fatty acids decrease LDL and cholesterol levels, while saturated acids increase both [18]. It has been found that n-3 and n-6 fatty acids expressed anti-inflammatory, antithrombotic, antiarrhythmic, hypolipidemic, and vasodilatory properties [19]. The n-3 fatty acids proved to be secondary preventers of several diseases and disorders such as coronary heart disease, hypertension, type 2 diabetes, renal disease, rheumatoid arthritis, ulcerative colitis, Crohn's disease, and chronic obstructive pulmonary disease [19] but also show effects in decreasing the risk of heart disease and cancer [20, 21].

Extraction represents the most common way of isolation of desired compound from the mixture, natural sources, or other matrices. Generally, extraction techniques may be divided into conventional and nonconventional. Usual conventional techniques are classic extraction and Soxhlet extraction. Such techniques usually applied toxic and environmental non-friendly solvents [22]. Due to such drawbacks, nonconventional techniques such as ultrasound-assisted, microwave-assisted, and subcritical water and supercritical fluid extractions have been developed.

Ultrasound (20 KHz–100 MHz) and microwave (300 MHz–300 GHz) penetrate through medium causing characteristic effects which enhance mass transfer. Ultrasound creates compression and expansion, thus producing cavitation [23]. Microwave on the other hand creates uniform heating as a consequence of medium resistance to ion flow. This effect is in close correlation with ionic conduction and dipole rotation mechanisms [23–25]. Supercritical water extraction relies on application of pressure to maintain water in liquid state during its heating above boiling point. Such state causes decrease in dielectric constant of water together with the polarity. Thus, polarity may, under certain conditions, become close to polarity of methanol [26–28].

Supercritical fluid extraction is a suitable substitution for conventional approaches such as hydrodistillation, steam distillation, and solvent extraction [29]. Techniques have started its commercial application during the 1980s [30]. Industrial scale of this technique found its application in processes such as decaffeination of green coffee beans and black tea leaves, production of hop extracts, isolation of essential oils, oleoresins and flavoring compounds from natural sources, extraction and fractionation of edible oils, and removal the pesticides from plant materials [31, 32]. This technique relies on application of fluids in their supercritical state. Such state can be achieved at pressures and temperatures above critical ones for given fluid. When this state has been achieved, fluid expressed properties between those characteristic for gas and liquid state [33, 34]. Density of fluid is similar to the density of liquids, while viscosity is similar to the values for gas. Behavior of two important factors influences the diffusivity which ranges between values for liquid and gas state. This results in better transport properties [34]. It should be mentioned that most important characteristic of fluids in this state is the ability to modulate their properties throughout modulation of density of fluid itself. This may be achieved by changing the pressure and temperature of system. Such changes in density influence directly the solubility of desired compounds in the fluid [35]. Carbon dioxide is the most commonly used fluid for this technique due to its nontoxic and nonexplosive properties, low price, availability, easy removable from extracts, and moderate critical properties ($T_c$ = 31.1°C; $p_c$ = 73.8 bar) [36].

Due to importance of fatty acids for proper function of human organism, it is also important to establish their occurrence in nature and possibility for utilization. The main goal of this research was to isolate and establish fatty acid profile in stinging nettle leaves. Different extraction techniques were used in order to compare their efficiency for isolation of fatty acids, while GC-FID technique was applied for their quantification. From the presented results, relationship between fatty acid structure and operational conditions was analyzed and established in the case of supercritical fluid extraction.

## 2. Experimental section

### 2.1. Plant material

Stinging nettle (*Urtica dioica* L.) leaves were purchased from Institute "Dr Josif Pančić" in 2016. Leaves were dried, grounded in blender, and stored in the paper bags until further processing.

### 2.2. Extraction procedures

Leaves were extracted using classic, Soxhlet, and supercritical fluid extraction methods. Classic extraction was performed by mixing 5.00 g of leaves with mixture of formaldehyde and ethanol (2:1; v/v) under vacuum at room temperature. Soxhlet extraction was performed using

5.00 g of leaves and 100.00 mL of petrol ether. Process was conducted in automated Soxhlet apparatus Soxtherm S306 (C. Gerhardt, Germany) at 150°C, under constant pressure for 2 h.

Supercritical fluid extraction was conducted using previously described apparatus (HPEP, NOVA-Swiss, Effretikon, Switzerland) [36, 37]. Main parts of this plant are gas cylinder with carbon dioxide, diaphragm-type compressor, extractor vessel with heating jacket, separator, pressure control valve, temperature regulation system, and regulation valves. Extractions were performed at three different pressure (100, 200, and 300 bar) and two different temperature (40 and 60°C) levels, while $CO_2$ flow rate was maintained constant (0.20 kg/h).

### 2.3. Sample preparation and fatty acid analysis

Extracts obtained using above-described techniques were further prepared for analysis using method described elsewhere [9]. Basically, fatty acids were extracted from prepared extracts using hexane, then hydrolyzed from their natural forms, and esterified using methanolic solution of KOH.

Analysis was conducted using Agilent 7890A gas chromatograph coupled with FID detector and CP-Sil 88 column (100 m × 0.25 mm × 0.20 μm), with nitrogen flow rate of 1 mL/min, applying previously described analytical method [9]. Nitrogen flow rate in detector was 25 mL/min, and air flow was 400 mL/min, while hydrogen flow rate was 30 mL/min. Oven temperature program was as follows: initial temperature 80°C (0.5 min), then linear increase of 4°C/min up to 220°C (4 min), and linear increase of 4°C/min up to 240°C (10 min). Temperatures of injector and detector were 240°C and 270°C, respectively. Standard solution for calibration was prepared by dissolving standard compounds in hexane in concentration range of 2–120 μg/mL. Limit of detection and recovery was 0.011–0.032% and 83–126%, respectively. Fatty acid contents were expressed as mg of fatty acid per g of extract (mg/g).

## 3. Results and discussion

### 3.1. Fatty acid contents in classic and Soxhlet extracts

Fatty acid profile of extracts obtained using classic and Soxhlet approaches is presented in **Table 1**, while chromatograms are given in **Figure 1**. According to their molecular structure, fatty acids (FAs) were divided into saturated (SFAs) and unsaturated (UFAs). UFAs were further divided into monounsaturated (MUFAs) and polyunsaturated (PUFAs) fatty acids.

Presented results indicated that automatic Soxhlet approach was a betted solution for isolation of fatty acids. Fatty acid yield in Soxhlet extract was 14.6-fold higher than yield of classic approach. Classic approach was proved to be better for isolation of UFAs, while SFAs dominated over UFAs in Soxhlet extract. Major FAs in classic extracts were C18:3 (3.368 mg/g), C16:0 (3.022 mg/g), C18:2 (2.775 mg/g), and C20:0 (1.036 mg/g). Domination of those three acids was earlier confirmed by several studies [9, 11, 12]. However, Soxhlet extract showed different profiles where C12:0 (45.639 mg/g) and C16:0 (40.523 mg/g) were major FAs followed by C18:1 (28.773 mg/g), 14:0 (20.656 mg/g), C18:0 (19.045 mg/g), C22:0 (13.385 mg/g), C20:0 (11.512 mg/g), and C24:0 (10.554 mg/g).

| Fatty acid | Content (mg/g) | |
|---|---|---|
| | Classic extraction | Soxhlet extraction |
| C6:0 | / | 0.667 |
| C8:0 | / | 4.925 |
| C10:0 | / | 4.677 |
| C12:0 | 0.202 | 45.639 |
| C14:0 | 0.456 | 20.656 |
| C14:1 | 0.085 | 0.610 |
| C15:0 | / | 0.722 |
| C15:1 | / | 0.375 |
| C16:0 | 3.022 | 40.523 |
| C16:1 | 0.158 | 0.548 |
| C17:0 | 0.054 | 0.568 |
| C17:1 | 0.072 | 0.288 |
| C18:0 | 0.493 | 19.045 |
| C18:1 *cis*-$\Delta^{9}$ | 0.486 | 28.773 |
| C18:2 *cis*-$\Delta^{9,12}$ | 2.775 | 6.579 |
| C20:0 | 1.036 | 11.512 |
| C18:3 *cis*-$\Delta^{9,12,15}$ | 3.368 | 2.232 |
| C21:0 | 0.137 | 1.075 |
| C20:2 *cis*-$\Delta^{11,14}$ | 0.423 | 1.663 |
| C22:0 | 0.890 | 13.385 |
| C20:4 *cis*-$\Delta^{5,8,11,14}$ | 0.087 | 0.935 |
| C23:0 | 0.174 | 0.400 |
| C22:2 *cis*-$\Delta^{13,16}$ | 0.218 | 0.775 |
| C24:0 | 0.644 | 10.554 |
| C20:5 *cis*-$\Delta^{5,8,11,14,17}$ | / | 0.685 |
| C24:1 | 0.251 | 1.096 |
| SFAs | 7.108 | 173.348 |
| UFAs | 7.923 | 44.559 |
| MUFAs | 1.052 | 31.690 |
| PUFAs | 6.871 | 12.869 |
| Total | 15.031 | 218.907 |
| SFA:UFA ratio | 0.90 | 3.91 |

**Table 1.** Fatty acid profile of classic and Soxhlet extracts.

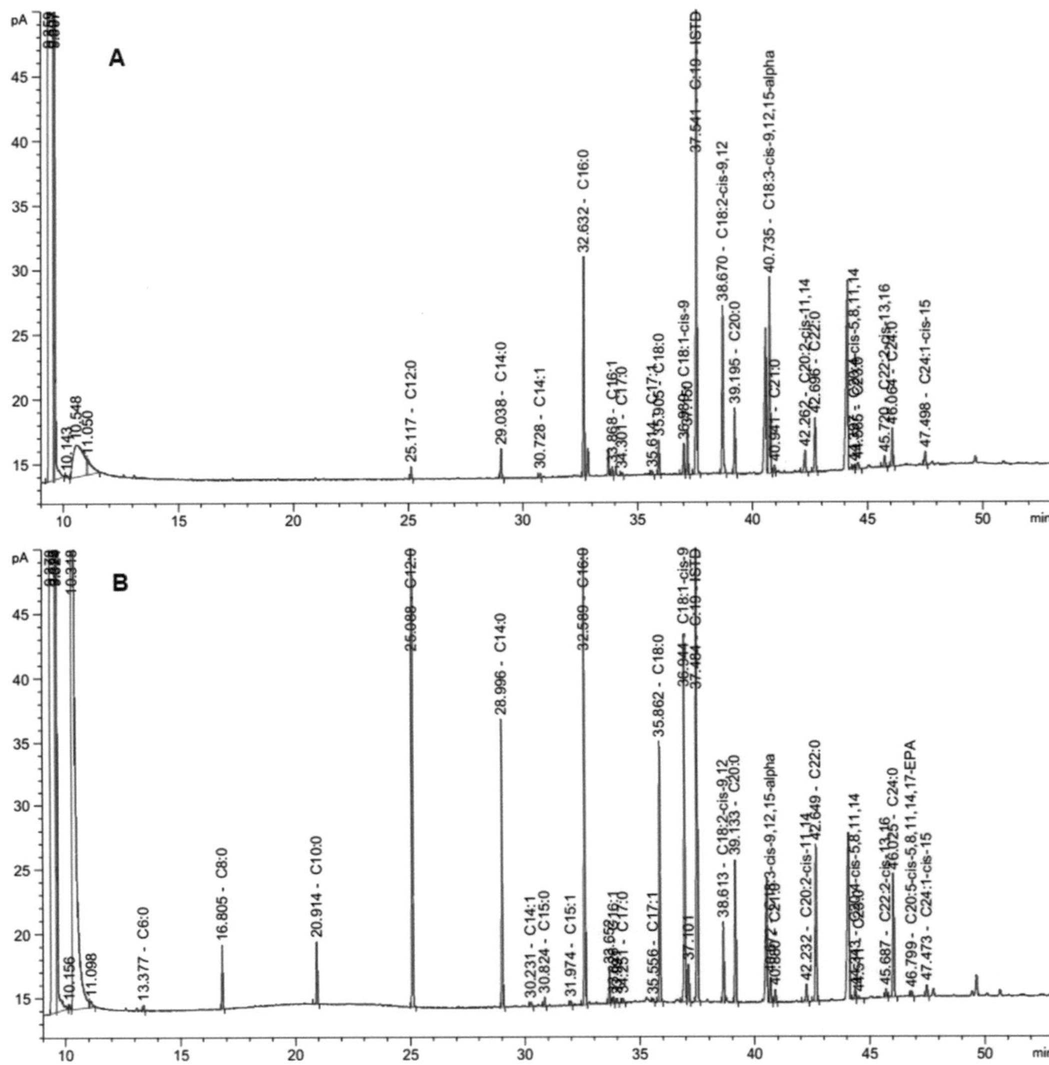


**Figure 1.** Chromatogram classic extract (A) and Soxhlet extract (B).

Besides the abovementioned differences, it might be also noticed that several fatty acids were detected only in Soxhlet extract (C6:0, C8:0, C10:0, C15:0, C15:1, and C20:5). Also, content of C18:3 PUFA was higher in classic extract. This might be explained with stronger interaction among C18:3 acid and ethanol, which was also previously reported [38].

### 3.2. Fatty acid profile of SFE extracts

Supercritical fluid extraction was conducted using three different levels of pressure and two different levels of temperature (**Table 2**).

Tuning these two parameters allows higher level of selectivity throughout influence on density of supercritical fluid and vapor pressure of solutes. Results of fatty acids GC analysis are presented in **Table 3**.

| Sample | Pressure (bar) | Temperature (°C) |
|---|---|---|
| 1 | 100 | 40 |
| 2 | 100 | 60 |
| 3 | 200 | 40 |
| 4 | 200 | 60 |
| 5 | 300 | 40 |
| 6 | 300 | 60 |

**Table 2.** Operational parameters of supercritical fluid extraction.

| Fatty acid | Sample/content (mg/g) | | | | | |
|---|---|---|---|---|---|---|
| | 1 | 2 | 3 | 4 | 5 | 6 |
| C6:0 | 0.376 | 0.213 | 0.219 | / | 0.208 | / |
| C10:0 | 0.190 | / | / | / | / | / |
| C11:0 | 0.128 | / | / | / | / | / |
| C12:0 | 0.209 | 0.216 | 0.178 | 0.158 | 0.175 | 0.228 |
| C13:0 | 0.211 | 0.181 | 0.178 | / | 0.143 | / |
| C14:0 | 3.859 | 1.697 | 2.150 | 1.573 | 1.968 | 1.523 |
| C14:1 | 1.155 | 0.860 | 0.659 | 0.414 | 0.611 | 0.483 |
| C15:0 | 0.204 | 0.151 | 0.173 | 0.404 | / | 0.165 |
| C15:1 | 0.185 | 0.142 | 0.612 | / | 0.376 | 0.338 |
| C16:0 | 8.657 | 6.616 | 8.010 | 7.102 | 6.053 | 6.331 |
| C16:1 | 1.075 | 0.588 | 0.432 | 0.702 | 0.497 | 0.556 |
| C17:0 | 0.350 | 0.294 | 0.148 | 0.763 | 0.192 | 0.235 |
| C17:1 | 0.553 | 0.239 | 0.329 | 0.250 | 0.161 | 0.171 |
| C18:0 | 1.996 | 1.527 | 1.799 | 2.279 | 1.388 | 1.631 |
| C18:1 *trans*-$\Delta^9$ | / | 0.126 | / | / | / | / |
| C18:1 *cis*-$\Delta^9$ | 2.130 | 1.565 | 1.640 | 1.072 | 1.426 | 1.297 |
| C18:2 *cis*-$\Delta^{9,12}$ | 10.900 | 7.746 | 9.047 | 7.347 | 7.216 | 7.240 |
| C20:0 | 3.851 | 3.986 | 7.734 | 13.934 | 4.286 | 10.107 |
| C18:3 *cis*-$\Delta^{6,9,12}$ | 0.431 | 0.276 | / | 0.160 | 0.266 | / |

| Fatty acid | Sample/content (mg/g) | | | | | |
|---|---|---|---|---|---|---|
| | 1 | 2 | 3 | 4 | 5 | 6 |
| C20:1 *cis*-$\Delta^{11}$ | 0.159 | 0.152 | / | 0.184 | 0.174 | / |
| C18:3 *cis*-$\Delta^{9,12,15}$ | 38.564 | 28.943 | 24.186 | 20.504 | 26.235 | 22.375 |
| C21:0 | 0.301 | / | 17.809 | 1.483 | 0.578 | 0.755 |
| C20:2 *cis*-$\Delta^{11,14}$ | 3.872 | 1.364 | 2.258 | 1.256 | 1.794 | 1.477 |
| C22:0 | 1.967 | 1.844 | 5.945 | 12.577 | 3.078 | 10.638 |
| C20:3 *cis*-$\Delta^{8,11,14}$ | 0.244 | 0.292 | 0.138 | 0.125 | 0.447 | 0.129 |
| C20:3 *cis*-$\Delta^{11,14,17}$ | / | 0.336 | / | / | / | 0.719 |
| C20:4 *cis*-$\Delta^{5,8,11,14}$ | 0.507 | 0.277 | / | 0.294 | 0.338 | 0.256 |
| C23:0 | 1.634 | 0.767 | / | 0.399 | 0.821 | 0.539 |
| C22:2 *cis*-$\Delta^{13,16}$ | / | / | 0.615 | / | / | / |
| C24:0 | 1.146 | 0.950 | 3.590 | 8.634 | 2.923 | 5.768 |
| C20:5 *cis*-$\Delta^{5,8,11,14,17}$ | / | 0.603 | / | / | 0.899 | / |
| C24:1 *cis*-$\Delta^{15}$ | / | / | 1.263 | / | / | / |
| C22:6 *cis*-$\Delta^{7,4,10,13,16,19}$ | 0.287 | 0.165 | / | 0.261 | 0.250 | / |
| SFAs | 25.079 | 18.442 | 47.933 | 49.306 | 21.813 | 37.920 |
| UFAs | 60.062 | 43.647 | 41.179 | 32.569 | 40.690 | 35.041 |
| MUFAs | 5.257 | 3.672 | 4.935 | 2.622 | 3.245 | 2.845 |
| PUFAs | 54.805 | 40.002 | 36.244 | 29.947 | 37.445 | 32.196 |
| Total | 85.141 | 62.116 | 89.112 | 81.875 | 62.503 | 72.961 |
| SFA/UFA ratio | 0.42 | 0.42 | 1.16 | 1.51 | 0.54 | 1.08 |

**Table 3.** Fatty acid profile of SFE extracts.

Pressure and temperature are considered to be the most influential parameters in SFE. Generally, pressure exhibited positive influence on extraction yield due to its positive influence on density of supercritical fluid. On the other hand, temperature exhibits dual effect as a result of combination of two variables: density and vapor pressure. Density decreases with temperature causing the decreasing in solubility. Vapor pressure increases with temperature, thus increasing the solubility in the same time [30, 39]. Pressure at the point of inversion of those two variables is known as crossover pressure or crossover point [30].

Influence of all abovementioned effects might be noticed in the case of fatty acid profile presented in **Table 3**. The highest yield of FAs was observed at 200 bar and 40°C (sample 3), while the lowest was at 100 bar and 60°C (sample 2). At both temperature levels, similar tendency of FAs yield might be noticed. Yields increased from 100 to 200 bar and then decreased at 300 bar. SFA yields showed same tendency, but the highest was observed in sample 4 (200 bar, 60°C), while the lowest was noticed in sample 2 (100 bar, 60°C). From the presented results, it

http://dx.doi.org/10.5772/intechopen.73533

was obvious that higher pressure and temperature levels were better for isolation of SFAs. In the case of UFAs, situation was completely different. The highest yield of UFAs was obtained in sample 1 (100 bar, 40°C), while the lowest was observed at 200 bar and 60°C (sample 4). It might be concluded that lower pressure and temperature levels were better for isolation of UFAs. This was also the same for both MUFAs and PUFAs. Such conclusions might be supported by SFA/UFA ratio, where the highest value was obtained in sample 4 (1.51) and the lowest in samples 1 and 2 (0.42). Chromatogram of sample 3 is presented in **Figure 2**.

Previous investigations of relationship between structure and solubility of organic compounds in supercritical carbon dioxide showed that length of carbon chain and number of double bonds influences their miscibility in supercritical fluids. Thus, increase in carbon number causes decrease in miscibility. Introduction of double bonds showed same effect, but unsaturation expressed favorable effect where alkenes were more miscible than alkanes [40]. Same trends stand for fatty acids. With increasing in number of carbon atoms and molecular weight miscibility decreases, while introduction of double bonds decrease melting point of fatty acids and this increases their miscibility [41, 42]. These parameters together with abovementioned influence of pressure and temperature ensured presented variations in fatty acid profiles (**Table 3**).

C6:0 acid was not quantified in samples 4 and 6. Comparing with its occurrence in samples 3 and 5, it might be concluded that increasing in temperature expressed negative effect on this acid. Same tendency was noticed in the case of C13:0 acid. C10:0 and C11:0 acids were quantified only in sample 1 (100 bar, 40°C). Negative influence of both pressure and temperature on miscibility of this fatty acid was clearly obvious. C15:0 acid was not quantified only in sample 5. C21:0 acid was not observed in sample 2, while C23:0 was not detected in sample 3. All other saturated acids (C12:0, C14:0, C16:0, C17:0 C18:0, C20:0, C22:0, and C24:0) were quantified in all samples.

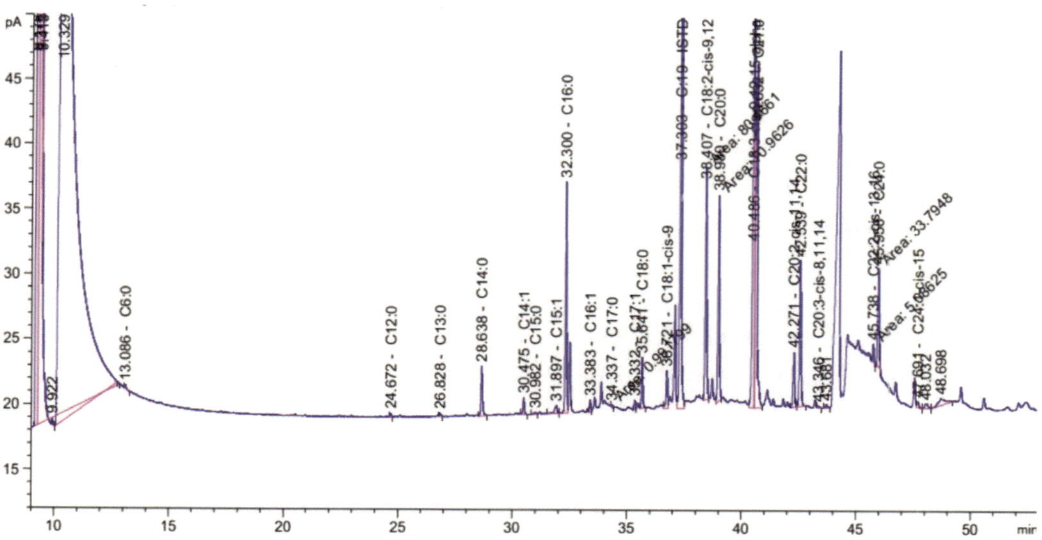


**Figure 2.** Chromatogram of sample 3 obtained at 200 bar and 40°C.

Comparing the yields for saturated FAs under isothermal conditions revealed existence of two trends: up to C18:0 acid yield decreased with pressure increasing at 40°C. After C18:0 acid, yield increased up to 200 bar and then decreased. On the other hand, at 60°C yield increased at first and then decreased. Changes in temperature under isobaric conditions influenced differently depending on fatty acid. At 100 bar yields decreased with increasing in temperature with the exception of C12:0, C20:0, and C23:0 where yield increased with the temperature. At 200 bar yields dropped except in the case of C15:0, C17:0, C18:0, C20:0, C22:0, C23:0, and C24:0. It might be noticed that more acids reached their maximal yield at higher temperature then in the case of 100 bar. At 300 bar yields decreased only in the case of C6:0, C13:0, C14:0, and C23:0. Such behavior might be explained with previously mentioned dual effect of temperature.

Effects of SFE conditions on yields of unsaturated acids will be considered separately for monounsaturated and polyunsaturated fatty acids. Results showed that *trans*-C18:1 acid was quantified only in sample 2, while 24:1 acid was quantified in sample 3. C15:1 acid was not detected in sample 4, while C20:1 acid was not quantified in samples 3 and 6. C14:1, C17:1, and *cis*-C18:1 acids were quantified in all samples. Isothermal conditions again showed dual effect. At 40°C yield constantly decreased with pressure increasing for C14:1, C17:1, and *cis*-C18:1 acids. In the case of isothermal process, at 60°C, yields reached their maximal values at 200 or 300 bar depending on acid. Isobaric conditions at 100 bar showed that yields dropped with increasing in temperature in all cases except for *trans*-C18:1 acid. At 200 bar trends were the same with the exception of C16:1 and C20:1 acid whose yields increased with temperature. In the case of 300 bar, yields of C16:1 and C17:1 decreased with increasing in temperature.

Polyunsaturated fatty acids were also detected and quantified in all samples. Dominant PUFA was C18:3 *cis*-$\Delta^{9,12,15}$ acid, which was the most abundant acid in all samples. C18:3 *cis*-$\Delta^{6,9,12}$ was not quantified in samples 3 and 6. C20:3 *cis*-$\Delta^{11,14,17}$ acid was detected only in samples 2 and 6, and C22:2 was found only in sample 3, while C20:5 was quantified only in sample 2. C20:4 was not found only in sample 3. Almost all PUFAs reached their maximal yields at 100 bar and 40°C (sample 1) with the exception of 20:3 *cis*-$\Delta^{8,11,14}$, 20:3 *cis*-$\Delta^{11,14,17}$, and C22:2 acids. They achieved their maximal yields in samples 5, 6, and 3, respectively. Both 20:3 acids achieved their maximal yield at 300 bar but at different temperature level. This might be explained with different positions of double bonds in their structures.

Deeper investigation showed that under isothermal conditions at 40°C, yields of most of compounds decreased up to 200 bar and then increased at 300 bar. Exceptions were C18:2 and 20:2 acids whose yields decreased at all three pressure levels. At 60°C, yields of C18:2 and C18:3 *cis*-$\Delta^{6,9,12}$ constantly decreased, while other acids showed different trends (yield increased or decreased up to 200 bar). Presented result showed that generally lower pressure was more beneficial for isolation of PUFAs and MUFAs (100 bar), while higher pressure is better for isolation of SFAs (200 bar). Temperature changes caused decreasing in yield of observed analytes, that is, UFAs, MUFAs, and PUFAs, at all three pressure levels (**Table 3**). They reached maximal levels at 40°C at 100 bar. Therefore, total yield of FAs showed same tendency where the highest yield was at 200 bar and 40°C followed by yield at 100 bat and 40°C. On the other hand, temperature negatively influenced SFA yield at 100 bar, but at 200 and 300 bars, yield of SFAs increased with temperature.

## 4. Conclusion

Three different extraction techniques were applied for isolation of fatty acids from leaves of stinging nettle. Results showed that automated Soxhlet approach provided the highest yield of fatty acids, while classic extraction achieved the lowest one. Comparison of those two conventional techniques revealed domination of certain saturated and unsaturated fatty acids. In the case of classic extraction, C16:0, C20:0, C18:2, and C18:3 were dominant compounds, while in the case of Soxhlet extraction, main acids were C12:0, C14:0, C16:0, C18:0, C18:1, C20:0, C22:0, and C24:0. It might be noticed that saturated acids absolutely dominated in this sample (SFA/UFA ratio was 3.91). Similar results were obtained in the case of SFE. Dominant acids were C16:0, C18:0, C18:1, C18:2, C20:0, C18:3, C21:0, C20:2, C22:0, and C24:0, but their content was in close connection with operational conditions, that is, pressure and temperature.

Investigation of influence of operational conditions on fatty acid profile revealed complex connection. The highest total yield was obtained in sample 3, followed by sample 1. Both samples achieved high yield at lower temperature level. Under isothermal condition pressure changes influenced differently. At both temperature levels, yield of SFAs increased up to 200 bar and then decreased. Yield of MUFAs decreased with pressure increasing at lower temperature but increase at higher temperature level and 300 bar. Yield of PUFAs decreased up to 200 bar and then increased at both temperature levels. Isobaric processes showed that temperature influenced negatively on yield of SFAs at 100 bar. At higher pressures, temperature exhibits positive effect, which was explained by changes in domination of two variables, that is, density and vapor pressure. On the other hand, temperature increasing negatively influenced yield of UFAs, MUFAs, and PUFAs. Results showed that lower pressure and temperature levels were most beneficial conditions for isolation of unsaturated fats. To conclude, careful choice of operational condition might result in favored isolation of desired group of fatty acids in this particular case.

## Author details

Saša Đurović[1,2]*, Saša Šorgić[3], Saša Popov[3], Marija Radojković[1] and Zoran Zeković[1]

*Address all correspondence to: sasatfns@uns.ac.rs

1 Faculty of Technology, University of Novi Sad, Novi Sad, Serbia

2 Institute of General and Physical Chemistry, Belgrade, Serbia

3 Oneological Laboratory, Vršac, Serbia

## References

[1] Orčić D, Francišković M, Bekvalac K, Svirčev E, Beara I, Lesjak M, et al. Quantitative determination of plant phenolics in *Urtica dioica* extracts by high-performance liquid

chromatography coupled with tandem mass spectrometric detection. Food Chemistry. 2014;**143**:48-53. DOI: 10.1016/j.foodchem.2013.07.097

[2] Upton R. Stinging nettles leaf (*Urtica dioica* L.): Extraordinary vegetable medicine. Journal of Herbal Medicine. 2013;**3**:9-38. DOI: 10.1016/j.hermed.2012.11.001

[3] Zeković Z, Cvetanović A, Švarc-Gajić J, Gorjanović S, Sužnjević D, Mašković P, et al. Chemical and biological screening of stinging nettle leaves extracts obtained by modern extraction techniques. Industrial Crops and Products. 2017;**108**:423-430. DOI: 10.1016/j.indcrop.2017.06.055

[4] Kukrić Z, Topalić-Trivunović L, Kukavica B, Matoš S, Pavičić S, Boroja M, et al. Characterization of antioxidant and antimicrobial activities of nettle leaves (*Urtica dioica* L.). Acta Periodica Technologica. 2012:257-272. DOI: 10.2298/APT1243257K

[5] Johnson TA, Sohn J, Inman WD, Bjeldanes LF, Rayburn K. Lipophilic stinging nettle extracts possess potent anti-inflammatory activity, are not cytotoxic and may be superior to traditional tinctures for treating inflammatory disorders. Phytomedicine. 2013;**20**:143-147. DOI: 10.1016/j.phymed.2012.09.016

[6] Gülçin İ, Küfrevioğlu Öİ, Oktay M, Büyükokuroğlu ME. Antioxidant, antimicrobial, antiulcer and analgesic activities of nettle (*Urtica dioica* L.). Journal of Ethnopharmacology. 2004;**90**:205-215. DOI: 10.1016/j.jep.2003.09.028

[7] Ghaima KK, Hashim NM, Ali SA. Antibacterial and antioxidant activities of ethyl acetate extract of nettle (*Urtica dioica*) and dandelion (*Taraxacum officinale*). Journal of Applied Pharmaceutical Science. 2013;**3**:96-99. DOI: 10.7324/JAPS.2013.3518

[8] Brown D. Encyclopedia of Herbs and their Uses. London: Dorling Kindersley; 1995

[9] Đurović S, Pavlić B, Šorgić S, Popov S, Savić S, Pertonijević M, et al. Chemical composition of stinging nettle leaves obtained by different analytical approaches. Journal of Functional Foods. 2017;**32**:18-26. DOI: 10.1016/j.jff.2017.02.019

[10] Gül S, Demirci B, Başer KHC, Akpulat HA, Aksu P. Chemical composition and in vitro cytotoxic, genotoxic effects of essential oil from *Urtica dioica* L. Bulletin of Environmental Contamination and Toxicology. 2012;**88**:666-671. DOI: 10.1007/s00128-012-0535-9

[11] Guil-Guerrero JL, Rebolloso-Fuentes MM, Isasa MET. Fatty acids and carotenoids from stinging nettle (*Urtica dioica* L.). Journal of Food Composition and Analysis. 2003;**16**:111-119. DOI: 10.1016/S0889-1575(02)00172-2

[12] Bağci E. Fatty acid composition of the aerial parts of Urtica Dioica (*Stinging nettle*) L. (*Urticaceae*). In: Şener B, editor. Biodiversity. Boston, MA: Springer US; 2002. pp. 323-327. DOI: 10.1007/978-1-4419-9242-0_40

[13] Pinelli P, Ieri F, Vignolini P, Bacci L, Baronti S, Romani A. Extraction and HPLC analysis of phenolic compounds in leaves, stalks, and textile fibers of *Urtica dioica* L. Journal of Agricultural and Food Chemistry. 2008;**56**:9127-9132. DOI: 10.1021/jf801552d

[14] Otles S, Yalcin B. Phenolic compounds analysis of root, stalk, and leaves of nettle. The Scientific World Journal. 2012;**2012**:1-12. DOI: 10.1100/2012/564367

[15] Sajfrtová M, Sovová H, Opletal L, Bártlová M. Near-critical extraction of β-sitosterol and scopoletin from stinging nettle roots. The Journal of Supercritical Fluids. 2005;**35**:111-118. DOI: 10.1016/j.supflu.2004.12.008

[16] Tack F, Verloo M. Metal contents in stinging nettle (*Urtica dioica* L.) as affected by soil characteristics. Science of Total Environment. 1996;**192**:31-39. DOI: 10.1016/0048-9697(96)05289-8

[17] Kara D. Evaluation of trace metal concentrations in some herbs and herbal teas by principal component analysis. Food Chemistry. 2009;**114**:347-354. DOI: 10.1016/j.foodchem.2008.09.054

[18] Lawrence GD. The Fats of Life: Essential Fatty Acids in Health and Disease. New Jersey: Rutgers University Press; 2010

[19] Simopoulos AP. Essential fatty acids in health and chronic disease. The American Journal of Clinical Nutrition. 1999;**70**:560S-569S

[20] Tapiero H, Nguyen Ba G, Couvreur P, Tew KD. Polyunsaturated fatty acids (PUFA) and eicosanoids in human health and pathologies. Biomedicine and Pharmacotherapy. 2002;**56**:215-222. DOI: 10.1016/S0753-3322(02)00193-2

[21] Iso H, Sato S, Umemura U, Kudo M, Koike K, Kitamura A, et al. Linoleic acid, other fatty acids, and the risk of stroke. Stroke. 2002;**33**:2086-2093

[22] Wang L, Weller CL. Recent advances in extraction of nutraceuticals from plants. Trends in Food Science and Technology. 2006;**17**:300-312. DOI: 10.1016/j.tifs.2005.12.004

[23] Azmir J, Zaidul ISM, Rahman MM, Sharif KM, Mohamed A, Sahena F, et al. Techniques for extraction of bioactive compounds from plant materials: A review. Journal of Food Engineering. 2013;**117**:426-436. DOI: 10.1016/j.jfoodeng.2013.01.014

[24] Letellier M, Budzinski H. Microwave assisted extraction of organic compounds. Analusis. 1999;**27**:259-270. DOI: 10.1051/analusis:1999116

[25] Jain T, Jain V, Pandey R, Vyas A, Shukla SS. Microwave assisted extraction for phytoconstituents—An overview. Asian Journal of Research in Chemistry. 2009;**2**:19-25

[26] Teo CC, Tan SN, Yong JWH, Hew CS, Ong ES. Pressurized hot water extraction (PHWE). Journal of Chromatography A. 2010;**1217**:2484-2494. DOI: 10.1016/j.chroma.2009.12.050

[27] Ko M-J, Cheigh C-I, Chung M-S. Relationship analysis between flavonoids structure and subcritical water extraction (SWE). Food Chemistry. 2014;**143**:147-155. DOI: 10.1016/j.foodchem.2013.07.104

[28] Anekpankul T, Goto M, Sasaki M, Pavasant P, Shotipruk A. Extraction of anti-cancer damnacanthal from roots of *Morinda citrifolia* by subcritical water. Separation and Purification Technology. 2007;**55**:343-349. DOI: 10.1016/j.seppur.2007.01.004

[29] Uquiche E, Huerta E, Sandoval A, del Valle JM. Effect of boldo (*Peumus boldus* M.) pretreatment on kinetics of supercritical $CO_2$ extraction of essential oil. Journal of Food Engineering. 2012;**109**:230-237. DOI: 10.1016/j.jfoodeng.2011.10.013

[30] Jesus SP, Meireles MAA. Supercritical fluid extraction: A global perspective of the fundamental concepts of this eco-friendly extraction technique. In: Chemat F, Vian MA, editors. Alternative Solvents for Natural Products Extraction, Berlin, Heidelberg: Springer Berlin Heidelberg; 2014. pp. 39-72. DOI: 10.1007/978-3-662-43628-8_3

[31] Brunner G. Gas Extraction: An Introduction to Fundamentals of Supercritical Fluids and the Application to Separation Processes, Topics in Physical Chemistry. Vol. 4. Heidelberg: Steinkopff; 1994. DOI: 10.1007/978-3-662-07380-3

[32] Brunner G. Supercritical fluids: Technology and application to food processing. Journal of Food Engineering. 2005;**67**:21-33. DOI: 10.1016/j.jfoodeng.2004.05.060

[33] Meullemiestre A, Breil C, Abert-Vian M, Chemat F. Modern Techniques and Solvents for the Extraction of Microbial Oils. Cham: Springer International Publishing; 2015. DOI: 10.1007/978-3-319-22717-7

[34] Herrero M, Cifuentes A, Ibanez E. Sub- and supercritical fluid extraction of functional ingredients from different natural sources: Plants, food-by-products, algae and microalgae: A review. Food Chemistry. 2006;**98**:136-148. DOI: 10.1016/j.foodchem.2005.05.058

[35] de Melo MMR, Silvestre AJD, Silva CM. Supercritical fluid extraction of vegetable matrices: Applications, trends and future perspectives of a convincing green technology. The Journal of Supercritical Fluids. 2014;**92**:115-176. DOI: 10.1016/j.supflu.2014.04.007

[36] Radojković M, Zeković Z, Mašković P, Vidović S, Mandić A, Mišan A, et al. Biological activities and chemical composition of *Morus* leaves extracts obtained by maceration and supercritical fluid extraction. The Journal of Supercritical Fluids. 2016;**117**:50-58. DOI: 10.1016/j.supflu.2016.05.004

[37] Zeković Z, Pavlić B, Cvetanović A, Đurović S. Supercritical fluid extraction of coriander seeds: Process optimization, chemical profile and antioxidant activity of lipid extracts. Industrial Crops and Products. 2016;**94**:353-362. DOI: 10.1016/j.indcrop.2016.09.008

[38] Péres VF, Saffi J, Melecchi MIS, Abad FC, de Assis Jacques R, Martinez MM, et al. Comparison of soxhlet, ultrasound-assisted and pressurized liquid extraction of terpenes, fatty acids and vitamin E from *Piper gaudichaudianum* Kunth. Journal of Chromatography A. 2006;**1105**:115-118. DOI: 10.1016/j.chroma.2005.07.113

[39] Paixao Coelho JA, Figueiredo Palavra AM. Supercritical fluid extraction of compounds from spices and herbs. In: Fornari T, Stateva RP, editors. High Pressure Fluid Technology for Green Food Processing. New York: Springer; 2015. pp. 357-396

[40] Dandge DK, Heller JP, Wilson KV. Structure solubility correlations: Organic compounds and dense carbon dioxide binary systems. Industrial and Engineering Chemistry Product Research and Development. 1985;**24**:162-166. DOI: 10.1021/i300017a030

[41] Yu Z-R, Singh B, Rizvi SSH, Zollweg JA. Solubilities of fatty acids, fatty acid esters, triglycerides, and fats and oils in supercritical carbon dioxide. The Journal of Supercritical Fluids. 1994;**7**:51-59. DOI: 10.1016/0896-8446(94)90006-X

[42] Güçlü-Üstündağ O, Temelli F. Correlating the solubility behavior of fatty acids, mono-, di-, and triglycerides, and fatty acid esters in supercritical carbon dioxide. Industrial and Engineering Chemistry Research. 2000;**39**:4756-4766. DOI: 10.1021/ie0001523